READING POPULAR PHYSICS

For my parents, Peter and Christine.

Reading Popular Physics

Disciplinary Skirmishes and Textual Strategies

ELIZABETH LEANE
University of Tasmania, Australia

ASHGATE

© Elizabeth Leane 2007

All rights reserved. No part of this publication may be reproduced, stored in a retrieval system or transmitted in any form or by any means, electronic, mechanical, photocopying, recording or otherwise without the prior permission of the publisher.

Elizabeth Leane has asserted her moral right under the Copyright, Designs and Patents Act, 1988, to be identified as the author of this work.

Published by
Ashgate Publishing Limited
Gower House
Croft Road
Aldershot
Hampshire GU11 3HR
England

Ashgate Publishing Company
Suite 420
101 Cherry Street
Burlington, VT 05401-4405
USA

Ashgate website: http://www.ashgate.com

British Library Cataloguing in Publication Data
Leane, Elizabeth
  Reading popular physics : disciplinary skirmishes and textual strategies
  1. Physics in literature 2. English literature – 20th century – History and criticism
  3. English literature – 19th century – History and criticism 4. Science in popular culture
  I. Title
  820.9'36'0904

Library of Congress Cataloging-in-Publication Data
Leane, Elizabeth.
  Reading popular physics : disciplinary skirmishes and textual strategies / by Elizabeth Leane.
     p. cm.
  Includes bibliographical references and index.
  ISBN-13: 978-0-7546-5850-4 (alk. paper)
  ISBN-10: 0-7546-5850-3 (alk. paper)
  1. Physics – Popular works. I. Title.
  QC24.5.L43 2007
  530–dc22
                                                                    2006032402

ISBN 978-0-7546-5850-4

Printed and bound in Great Britain by MPG Books Ltd, Bodmin, Cornwall.

Contents

List of Tables   *vii*
Acknowledgements   *ix*

Introduction   1

1   Popular Physics Books: A Brief History   19

2   "I blame the popularisers …": The Boom and Its Backlash   41

3   The "Two Cultures": Some Theoretical Developments   61

4   Knowing Quanta: Anthropomorphic Metaphor in Popularizations of Quantum Theory   81

5   Exploding the Big Bang: Popular Cosmology as Mythic Narrative   107

6   *Chaos, Complexity* and Characterization: Stereotypes of the Scientist in Physics Popularizations   137

Conclusion   163

Bibliography   *167*
Index   *191*

List of Tables

2.1 Bestselling Popular Science Books, 1978–1979   44
2.2 Bestselling Popular Science Books, 1996–1999   45

Acknowledgements

There are many people I would like to thank for their contributions to the development of this book.

Encouragement, support and advice from staff and students at several universities have been vital to my research. Phil Waldron at the University of Adelaide provided inspiration at an early stage in my academic career. Heath O'Connell was there at the beginning of my encounter with physics, and his friendship and interest remain important to me. At the University of Oxford, regular lunchtime meetings with Noor Al-Abood, Subarno Chattarji, Sarah Broom, Diana Lewis and Georgina Taylor kept me sane, and the friendship of Roger Arjoon, Mark Cassidy, Mark Jenkinson, Cam Keppert, Lori Ormrod, Peter Rutledge, Annette Salmeen, Heidi Stalla and Dale Wilson was (and is) invaluable. Michael Whitworth pointed me to J. W. N. Sullivan's articles and other helpful resources. My colleagues at the University of Tasmania have been supportive and helpful, particularly Ian Buchanan, Ralph Crane, Lisa Fletcher, Lucy Frost, Samantha Hardy, Anna Johnston, Jenna Mead and Julia Petzl. Special thanks go to my unflagging research assistant, Stephanie Pfennigwerth, for her hard work, sharp eye for detail and constant encouragement. I would like to thank Ashgate's two anonymous readers, as well as John Carey, John Durant and particularly David Bradshaw for their criticism and advice on earlier versions of the manuscript. Librarians at the Bodleian and Morris Miller libraries were constantly helpful in tracking down material. I am also very grateful to the Rhodes Trust for supporting the Oxford component of the research.

A number of science writers, journalists and publishers kindly granted me interviews and provided me with information and advice. I would like to thank Rosalind Arden; Keith Clack of Blackwell's Bookshop, Oxford; Paul Davies; Roger Highfield of *The Daily Telegraph*; Tom Holman, Stephen Butler and Peter Harland of Bookwatch; Ravi Mirchandani of Penguin Books; Tim Radford and Robin McKie of *The Guardian*; Will Sulkin of Random House; Susan Wakefield of Whitaker BookTrack; and Robyn Williams of the ABC Radio Science Unit. Paul Davies, whose books I devoured as a teenager and whose class I took as an undergraduate at the University of Adelaide, has been particularly encouraging and generous with his time.

Ashgate have made the publication process a smooth and enjoyable experience. In particular, I would like to acknowledge Ann Donahue, Anne Keirby and Melissa Pearce.

Lastly, I would like to thank my family, especially Damian, my favourite physicist and my mainstay; my brothers, Matt and Josh, who originally inspired me to study science; and my parents, Christine and Peter, who taught me the value of clear thinking and persistence, and provided practical, intellectual and personal help throughout this project. All of their unconditional love, encouragement and support made this book possible.

Different versions of the material in Chapters 4, 5 and 6 of this book have been previously published. An earlier version of Chapter 4 was published as "Knowing Quanta: The Ambiguous Metaphors of Popular Physics" in *The Review of English Studies*, 2001, volume 52, issue 207, pp. 411–31, and is reprinted here with permission of Oxford University Press. Part of Chapter 5 appeared in *Crossing Boundaries: Thinking Through Literature*, ed. Julie Scanlon and Amy Waste, Sheffield Academic Press, 2001, and is reprinted here with permission of the Continuum International Publishing Group. Part of Chapter 6 was originally published in *Essays in Arts and Sciences*, volume 32, and is reprinted here with permission of Essays in Arts and Sciences, University of New Haven. Quotations from *The Dancing Wu Li Masters* by Gary Zukav, published by Rider, are reprinted by permission of The Random House Group Ltd.

Introduction

> What with their tradition of nerdiness and relish for vivisection, scientists used to be anything but popular. But now, spurred on by canny publishers and sinister drug companies out to improve their public image, scientists have staged a strange little comeback. Ten years since Hawking's book—that incomprehensible tome on time—and now it's considered *normal* for science books to fill the bestseller lists, and for their authors to appear on TV. (Ellmann 2)[1]

It is close to half a century since C. P. Snow introduced his notion of the "two cultures," and many commentators now see his argument as outdated and redundant. However, the above quotation from an article by novelist Lucy Ellmann, published in the London *Guardian* newspaper in mid-1998, suggests otherwise. Snow argued that literary intellectuals "like to pretend that the traditional culture is the whole of 'culture'" (14); Ellmann is alarmed that science books vie with fiction on the bestseller lists, and defensively downgrades science's popularity as a "strange little comeback." Snow claimed that the "total incomprehension" of science gave "an unscientific flavour to the whole 'traditional' culture," a flavour often "on the point of turning anti-scientific" (11); Ellmann dismisses Stephen Hawking's bestseller as an "incomprehensible tome," and her assertion that scientists have a "tradition of nerdiness and relish for vivisection" is blatantly hostile. There is one aspect of Ellmann's attack, however, which Snow did not anticipate: her hostility towards science is focused directly on science *popularizations*, which she believes are enjoying unprecedented success.

The purpose of this book is to investigate why, and how, popular science books (specifically, popular physics books) have played such an important role in recent exchanges—some hostile, others fruitful—between literature and science. There are two main contexts for this investigation. The first is the "popular science boom"—Ellmann's "strange little comeback"—which began rumbling in the late 1970s and reached its peak in the 1990s. This publishing phenomenon saw popular science writing lauded as a genre on equal footing with literature, but it simultaneously catalysed lingering "two cultures" hostilities into action, as epitomized by Ellmann's attack. The second context is the "Science Wars": the challenges to the epistemological and social status of science emerging from post-structuralist philosophy, sociology, literary criticism and cultural studies, and the hostile response they generated in some parts of the scientific community. Popularizations of science were one forum in which these "wars" were waged.

For a literary critic, an obvious starting point for an investigation of popular science books is the interaction between popularizations and literature. That such an interaction is taking place is not in doubt. In her article Ellmann attacks the traitors

---

1 Throughout this book, italics in quotations are original unless otherwise stated.

to literary culture who have welcomed science popularizers into their territory: Melvyn Bragg who "spread it a bit by inviting scientific dullards on to his BBC radio show," and the "male novelists of England" who "quickly fell into line, convinced they should not write about life, the world, the Universe, their mothers, or anything else until they'd read a few books on black holes" (2). Six years earlier, a similar point was made in a very different forum, a satirical cartoon in the popular science magazine *New Scientist*. The cartoon's protagonist is dismayed to find that his "juicy new novel" is packed with scientific references (mostly drawn from physics):

> Science, Science, Science! Day in, day out, I grapple with it ... Newts—neutrinos—nebulae—Newton's rotten rings ... A bit of escapism!—Is it too much to ask?—They're all at it, you know ... it's not just science fiction if it's Martin Amis, oh no ... practically give him the Booker Prize for it, don't they? ... & what would they be doing if they hadn't heard about fancy science?—They'd have to be thinking up stories! Tales! I blame the popularisers ... . (Charlesworth)

It is not difficult to find evidence for these observations. Even ignoring science fiction, where such borrowings might be expected, it is possible to list numerous novelists and playwrights who draw on popularizations in their literary texts. Besides Amis, who borrows from the content and characteristic discourse of popularizations in such novels as *London Fields* and *Night Train*, one could mention Tom Stoppard, whose *Hapgood*, a play about quantum mechanics, begins with a quotation from a popular physics book, Richard Feynman's *The Character of Physical Law*, Michael Frayn, whose *Copenhagen* requires its actors to perform a popularization of quantum mechanics on stage; or Ian McEwan, who acknowledges physicist David Bohm's *Wholeness and the Implicate Order* as a source for *The Child in Time*, and has a science writer as the protagonist of his *Enduring Love*. In contrast to Ellmann's assertion, not all of these writers are male or English. Consider Margaret Atwood, who in the acknowledgements to her novel *Cat's Eye* states, "The physics and cosmology sideswiped herein are indebted to Paul Davies, Carl Sagan, John Gribbin and Stephen W. Hawking, for their entrancing books on these subjects." In her more recent novel, *Oryx and Crake*, Atwood thanks the many "non-fiction science writers" who provided her with "[d]eep background" (435). Janet Turner Hospital lists three physics popularizations, including Fritjof Capra's *The Tao of Physics*, as "helpful background material" for her novel *Charades*. As their acknowledgements indicate, all of these authors incorporate into their novels concepts and sometimes styles borrowed from science popularizations.

The point here is not simply that these novelists have been influenced by ideas from science—this would not be a new, unusual, or unstudied phenomenon. As literary critic Lance Schachterle points out, "One sign of the inadequacy of C. P. Snow's thesis of 'The Two Cultures' is how frequently present-day writers turn to contemporary physics for underlying metaphors." Mentioning a number of writers, including Nabokov, Fowles, Barth, Updike, Vonnegut, Pynchon and DeLillo, Schachterle notes, "How these authors have learned enough about modern physics to use scientific metaphors powerfully is unknown; traditional influence studies, available for earlier periods, remain to be done for these post-World War II writers" (177). What characterizes the writers I have mentioned is that they explicitly identify their sources as popular science books. Their novels are situated not simply within a culture that is informed by scientific discourse, but rather a culture informed specifically and wholly by *popular* scientific discourse. These writers are not structuring their fiction around any first-hand knowledge of physics, as might be imagined in the case of, say, Pynchon. Rather, they are self-consciously employing knowledge of and interest in *popularizations* of science, and, in some cases, relying on the fact that these popularizations are becoming so widespread and well known that readers will readily understand their meaning.

Novels and plays such as these, then, would provide a clear point of entry for the literary critic interested in the role of popularizations in cross-disciplinary exchange. However, an analysis of novels, plays, or poetry would in some ways merely reinscribe a traditional division between literature and science: the belief that literary language is open to analysis while the language of science provides a transparent window on physical reality. Many contemporary critics have pointed to the inadequacy of a project that concentrates solely on the representation of science in literature. Stuart Peterfreund, in his introduction to the collection of essays *Literature and Science: Theory and Practice*, complains that "the study of literature and science has all too often been, especially prior to the 1980s, a matter of 'finding' scientific ideas 'in' literature and literary ideas 'in' science." This kind of work, Peterfreund argues, "is important for establishing the discursive space necessary for 'doing' literature and science," but "never did—indeed, never could—venture meaningful commentary on the foundational assumptions and methodologies used to study literature and science or on the ideological dimension of the resulting discourse" (5). Because I do want to venture such meaningful commentary, I have chosen to focus directly on the language—the "textual strategies"—of physics popularizations themselves. Such an analysis reveals the important role these books can play as sites of exchange between literature and science: places where disciplinary skirmishes are covertly fought, but also places which offer the possibility of looking beyond the worn-out "two cultures" model.

Popular Science and the "Two Cultures"

My claim that popular science books can be usefully viewed as an interface between the "two cultures" rests on the assumption that all cross-disciplinary exchange (including the "Science Wars") involves popularization. According to media studies researcher Jose van Dijck, Snow's argument implies "the dire need to translate between expert and lay communities ... . His concern for scientific illiteracy and unbridgeable gaps was not restricted to academic communities of arts and humanities scholars" (181). Any closing of the "two cultures" gap must indeed involve communication between expert and lay communities, but one should not equate (as van Dijck effectively does) scientists with the former and "arts and humanities" researchers with the latter. Each expert group becomes a "lay communit[y]" with respect to the other, and Snow was concerned about illiteracy in literature as well as in science. Any exchange of discipline-based information between these two elites will necessarily be a popularization, in that it is a communication from an expert to a reader outside of his/her specialist field. With very few exceptions (presumably Snow would have considered himself one), the literary community has access to scientific ideas only through popularizations of one form or another, and vice versa. Thus in a basic sense popularizations mediate between the literary and scientific communities in that they allow exchange of information: this applies to popularizations of both science and literature, but it is the former on which I am concentrating here.

Popular science books, however, do not only explain a particular concept or field within science. They can also act as forums for scientists to promote or defend rival views within a scientific field, or to engage, implicitly or explicitly, in cross-disciplinary debate. Richard Dawkins's *River Out of Eden*, for example, is primarily an explication of evolutionary biology, but Dawkins pauses at one point to defend evolutionary theory against "cultural relativism" (35), noting that "If you are flying to an international congress of anthropologists or literary critics, the reason you probably get there ... is that a lot of Western scientifically trained engineers have got their sums right" (36). This kind of positioning of science's role and status in society is perhaps better understood as "public science" rather than popularization, but popularizations are often the sites where "public science" takes place. Popularizations thus potentially mediate between science and literature (or other non-scientific disciplines) in a more complex, and often more direct, manner than simply through familiarizing non-scientific readers with scientific ideas.

Popularizations also mediate between science and literature in a third sense: writing popularizations, scientists come into immediate contact with the tools of the literary trade. Admittedly, not all popularizers adopt literary modes of expression with the glee of physicist David Mermin, who, casting about for ways to popularize Einstein's Special Theory of Relativity, settled upon a "one-act relativistic tragicomedy set in otherwise empty space" (*Boojums* xvii, 207). Nevertheless, some form of appropriation of literary technique is inevitable; for, unlike the writer of professional scientific discourse, who can assume his/her reader's interest, the popular science writer must capture the reader's attention and imagination just as a novelist must, and to do this, s/he must borrow from the novelist's tool-kit. Several literary critics have brought attention to this point. Gillian Beer observes, "When they are writing outside of the tight circle of fellow professionals, the best scientific communicators excel by using the possibilities of current literature" (*Open Fields* 170). Elinor Shaffer in her introduction to the anthology *The Third Culture: Literature and Science* suggests, "Popular scientific writing brings into play a range of fictional techniques that the literary analyst may be well placed to recognize" (4). Robert Kelley writes, "While the traditional scientific paper seems devoid of the rhetorical play and obvious fictionality of the novel, the semipopular science text lies somewhere in between traditional scientific and novelistic texts" (133). And most prominently, John Carey in his introduction to *The Faber Book of Science* (1995), published at the peak of the boom, claims popular science writing as "a new kind of twentieth-century literature" (xiv). Popular science thus mediates between science and literature by presenting the content of the former through some of the established techniques of the latter.

Once the literary nature of popular science is recognized, literary analysis of this genre appears not only possible but indispensable. In a 1995 review of Carey's anthology, John Durant, then Professor of the Public Understanding of Science at Imperial College, observed that "for too long, literary circles have all but ignored scientific writing" and called for "a literary criticism of popular science writing." Literary critics have been more alert to popular science writing than Durant suggests; however, it is true that the late twentieth-century boom solidified a sense of the genre as both culturally significant and worthy of serious literary analysis. *Reading Popular Physics* is part of the renewed literary interest in popular science that Carey and Durant identify and encourage. It draws together and examines the ways in which popular science acts as a kind of "bridge" over the "two cultures" chasm, and also its potential to provoke a widening of the chasm (as Ellmann's article amply illustrates).

Why Popular Physics Books?

"Science popularization" is an enormous field: it spans not only several centuries and a proliferating number of disciplines and sub-disciplines, but also a highly diverse range of media and activities. It encompasses thousands of books, essays, magazines, newspaper articles and—depending on the breadth of one's definition of "popularization"—textbooks and other educational material, grant proposals, government reports, not to mention non-print material such as television and radio programmes, websites and electronic publications, public lectures, science museums, festivals and exhibitions. While for convenience these diverse genres are grouped under the umbrella of "science popularization," each deserves (and in many cases has received) its own analysis, drawing on special expertise.

Reading Popular Physics focuses on a medium of science popularization that the literary critic is well placed to examine: books. While the popular science boom of the late twentieth century is most often described by commentators as a boom in *books* and *book sales*, it was not of course confined to this medium.[2] There are, however, several reasons why books are an important focus of analysis. Bruce Lewenstein, in an article entitled "How Science Books Drive Public Discussion" (2002), warns against the dangers of focusing research in science communication only on new media such as the World Wide Web, and makes a case for the prominent role of recent popular science books (such as *A Brief History of Time*) within public debate (75). He argues that books are "where we traditionally turn for culture," and hence popular science books "show the integration of science and culture in our everyday lives" (73). Other researchers note that books have an impact that reaches

---

2 For instance, *Sight and Sound* magazine in 1998 reported that "The popularisation of science on television continues, taking its cue from recent trends in publishing ... science now seems to be the cultural centre of 90s programming" (Cathode). See Fahnestock ("Accommodating Science" 18) for a summary of a boom in popular science magazines from the 1970s onwards, and Lewenstein ("Was There Really a Popular Science 'Boom'?"; "Why Isn't Popular Science More Popular?") for an analysis of the success of popular science in newspapers, magazines and television from the late 1970s to the mid-1980s.

further than their direct readerships, as they often provoke debate, discussion and further popularization in the press (Cassidy 126–7; Dixon 379). Lastly, it is books, rather than websites, magazines or television, that are associated most readily with literature and the literary community. For a project that seeks to examine the way that popularization mediates between literature and science, books are an apposite focus. It is thus popular science books published in the boom years (defined here as approximately the last quarter of the twentieth century) on which this book concentrates. Most of my examples are texts by British or North American authors. Although in later chapters I suggest that the cross-disciplinary debate takes different forms in these two regions, many recent science popularizations are published and distributed internationally, and I have emphasized the nationality of writers only where this is relevant to their work.

I discuss popular books dealing with a range of scientific disciplines throughout this study, but my close focus is on physics books. My first degree is in mathematical and theoretical physics, and although I would not suggest that a literary critic must or should have scientific training to conduct this kind of analysis, I also believe that critics who do have this background bring a useful and unusual combination of skills and knowledge to the task. "Physics" carries a set of associations that make it both particularly problematic and particularly fruitful in the context of this study. There has been a strong tendency in the past, among both scientists and non-scientists, to endorse a hierarchy of sciences which privileges physics above chemistry, physical sciences above biological sciences, and biological sciences above social sciences. This hierarchy was first outlined by the nineteenth-century positivist philosopher Auguste Comte, who based his scale of "relative perfection of the different sciences" on the observation that "the more general, simple, and abstract any phenomena are, the less they depend on others, and the more precise they are in themselves," and hence the greater the "possibility of applying mathematical analysis" to their study (1: 29). Two main characteristics of physics thus underlie its perceived pre-eminence. The first is its highly mathematical and rigorous nature, which is the basis for its status as the "purest" or "hardest" of sciences, "a kind of gold standard against which weaker or debased forms of science could be measured" (Collini xlvii). The second is its focus on the most "basic" or "fundamental" natural phenomena, such as time, space and matter, which leads to the unabashedly reductionist claim that physics somehow encompasses all other sciences. Ernest Rutherford's famous dismissal of all scientific disciplines except physics as equivalent to "stamp collecting" is a notorious example of this attitude (qtd. in Blackett 108). Today's practitioners, while employing more circumspect language, often reinforce this image: "Physics is the most pretentious of the sciences," writes physicist-popularizer Paul Davies, "for it purports to address all of physical reality. The physicist may confess ignorance about a particular system—a snowflake, a living organism, a weather pattern—but he will never concede that it lies outside the domain of physics in principle. ... the entire universe, from the smallest fragment of matter to the largest assemblage of galaxies, becomes the physicist's domain ..." ("The New Physics" 1).

This privileging of physics over other sciences can result in what is sometimes termed "physics envy": the tendency for traditionally "softer" sciences to desire or attempt, often inappropriately, to emulate the mathematical rigour of physics.

As Sal Restivo observes, in the area of science studies, this hierarchization can result in a corresponding privileging of analyses that focus on physics: "While the contemporary picture is not as stark as the traditional one, studying physicists and Nobel prize winners still seems to bring easier and more widespread recognition than studying agricultural scientists, physicians, engineers, and everyday 'puzzle-solvers' ..." ("Science Studies" 14). Restivo concludes that science studies researchers need to stop equating science with physics, thereby demoting other sciences to "second-class modes of inquiry" (15).

While I would not want this book's focus on physics to endorse such a hierarchy, the traditional *image* of physics as the most objective, rigorous and "pure" science is undeniably important to this discussion, as it is implicitly constructed in opposition to the traditionally "softest," most subjective discipline within the humanities: the study of literature. It is no coincidence that Snow, himself a physicist and a novelist, first defined his two cultures not as the humanities and the sciences but more specifically as "literary intellectuals" and "scientists, and as the most representative, the physical scientists" (4). According to Snow, the "physicists" and the "literary intellectuals" occupy the extreme poles of academic society, although there might be "all kinds of tones of feeling" in between (10-11). For an analysis that is interested in the role of popular science books in mediating between the "two cultures," physics is a particularly useful focus.

Furthermore, popular physics books have played a key role in twentieth-century science popularization, both in terms of sheer numbers of publications and in determining how the relationship between science and "the public" is structured. On the first point, Roger Smith, in his bibliography of popular physics, states that "astronomy and popular physics dominated science sections in bookstores and public libraries" throughout the twentieth century (1). The most prominent popular science book of the late twentieth century, Hawking's *A Brief History of Time*, was a physics book, as was the outstanding science bestseller of the early decades of the century, James Jeans's *The Mysterious Universe*. On the second point, historian of science Bernadette Bensaude-Vincent argues that "[t]he notion of an increasing gap between science and the public is heavily dependent upon twentieth-century physics" ("A Genealogy" 109). The "new physics," she claims, resulted in a shift away from earlier understandings of science as a form of common sense and "public reasoning" to something far more abstruse and mathematical (107, 109). Contemporary physics, then, is perceived as less accessible than any other science, and the challenges of popularizing it more marked. Popular physics is thus of central importance to the study of twentieth-century science popularization.

Defining "Popular Physics": Some Problems and Approaches

"Popular science" and "popular physics" are not easily defined terms. There are a host of competing and overlapping phrases which might be used to describe communication between scientific specialists and their non-specialist audiences, all with specific histories and connotations within research: popular science, popularization of science, pop science, expository science, the literature of science, science communication, public science, and public understanding of science. The multi-disciplinary nature of recent analysis in this field, and changing conceptualizations of popularization, mean that my choice of terms needs careful explanation.

One source of confusion between terms is the looseness and historical instability of language in general. "Popular science," for example, could be intended to mean "science of the people," or "science in popular culture," or "popularized science." Bensaude-Vincent observes that before the twentieth century, "popular science" suggested "a science distinct from that of professional scientists" ("In the Name of Science" 322), but now is "not supposed to refer to any thing other than the image of science as reflected by vehicles of pop culture such as commercial advertisements, best-seller fictions or television serials" (322). Thus for Bensaude-Vincent, "popular science" is distinct (in both periods) from "popularised" science or "science popularisation" (321–2). However, George Basalla, another historian of science, states that "the portrayal of science in popular culture is not to be confused with what is generally called 'popular science,'" and to avoid confusion he terms the former "pop science" (261). In my study I follow Basalla in distinguishing popular science from the wider category of the representation of science in popular culture (which would include, for example, most science fiction).

A second kind of terminological ambiguity occurs when "popular science" overlaps with other categories such as "public science" and "public understanding of science." Both of these categories involve science popularization and are occasionally conflated with it. According to historian of science Frank Turner, the term "public science" refers to "the body of rhetoric, argument, and polemic" produced by scientists in the process of "justify[ing] their activities to the political powers and other social institutions," and these scientists are "public scientists" (589); it is this definition of public science that I employ here. Defined in this way, public science overlaps but is not identical with popular science;[3] however, critics sometimes appear to use the two terms interchangeably. For Lewenstein, a science communication researcher, "public science" signifies a different idea again, indicating occasions when science is drawn on in discussions of general-interest topics, such as sex, psychological well-being or exploration ("How Science Books" 72). "Public understanding of science" (PUS) is a term which, confusingly, is variously used to describe a state of affairs, an academic field, and an institutionally sponsored movement. Although PUS concerns are centuries old, the late twentieth century saw heightened attention to PUS issues, including (in Britain) the commissioning of a report by the Royal Society in 1985, the forming in 1986 of COPUS (the Committee on the Public Understanding of Science), the establishment of the Science Book Prizes in 1988 (now the Aventis Prizes) and the launching of the academic journal *Public Understanding of Science* in 1992. Central to PUS initiatives was the fostering of high-quality popularizations, as epitomized by the Book Prizes.

---

3   Roger Cooter and Stephen Pumfrey note that accounts of public science "show how (usually élite) scientists mobilized (usually élite) resources in order to win assent from (usually élite) audiences," and tend not to focus on "ordinary men and women" (245); the concept of "popular science" implies a wider variety of audiences/readerships.

However, activities categorized under PUS imply a social and moral imperative that is not necessarily present in science popularization, and the distinction between the two should be maintained.[4]

A third source of confusion in terminology is a shift in the conceptualization of science popularization and the approach to its study. It is now commonplace for analyses of popular science within a range of disciplines to begin by distancing their approach from the "traditional" model of science popularization. The central concern expressed by critics is that "popularization" of science has traditionally been viewed by researchers (as well as many practitioners) as a flow of established, unalterable knowledge from an authoritarian scientific community into a passive public. In this hierarchical model, the audiences for popular science are considered to be "atomistic receivers of information, with little or no collective internal structure, who passively internalise knowledge in isolation from other social activities and structures" (Whitley 4). The producers of scientific knowledge, in contrast, are conceived as "a highly organised community with clear boundaries who use their esoteric and elaborate skills to generate 'true' knowledge in isolation from non-scientists" (Whitley 5–6).

Since the mid-1980s, this "traditional" model of science popularization has been increasingly questioned and challenged by a number of critics on a variety of counts.[5] The model assumes a bipolar structure of "professional" versus "popular" science, or "science" and the "public," categories that are themselves internally complex, and might be better considered as two ends of a continuum. It neglects the heterogeneous nature of audiences by positing all targets of popularization as those with little or no scientific education, and therefore limiting the understood range of exchanges that might occur between specialists and non-specialists. It separates popularization from the "knowledge production and validation process" (Whitley 3) and leaves no room for the possibility of feedback between popularization and scientific research. Popularizations are seen as translations—or, if unsuccessful, distortions—of scientific knowledge, rather than producers of knowledge in their own right. As a consequence, the traditional model fails to recognize the possibility of a separate form of scientific knowledge within popular culture which is not necessarily a reflection of professional science, and may even oppose the latter.[6] Conversely, the traditional view of popularization is based on an outdated understanding of science itself: it ignores challenges to the image of scientific knowledge as the product of "cohesive autonomous communities" (Whitley 6), and reinforces the traditional conception of professional science as elitist, objective, and culturally and discursively autonomous: "A concept of purity requires one of contamination, and the notion of popularization shores up an idealized view of genuine, objective, scientifically-certified knowledge" (Hilgartner

---

4   See Gregory and Miller for a discussion of science popularization in relation to PUS.

5   See Shinn and Whitley, Preface; Whitley; Cloître and Shinn; Hilgartner; Cooter and Pumfrey; Myers, "Discourse Studies."

6   See Sheets-Pyenson 551; Bensaude-Vincent, "In the Name of Science" 321–2; Cooter and Pumfrey 248–9.

520). These criticisms reflect the growing influence of constructivist theories of scientific knowledge (discussed in Chapter 3) which conceive of science as a socially and culturally embedded activity rather than something self-contained and isolated.

In place of this traditional model, critics argue for a more flexible model that recognizes the diversity of exchanges that might be classed under the category "science popularization" and does not judge popularizations only in terms of how accurately they convey concepts or ideas of non-popular (i.e. professional) science. This wider conception of the process of popularization allows for a more complex understanding of the composition of audiences and readerships, acknowledges the variety of exchanges that can occur between a scientific specialist and non-specialist, sees these exchanges as two-way rather than one-way movements of information, and hence recognizes popularizations as part of a system of knowledge production, rather than separate from it.

This shift in understanding of the popularization process has led to debates about terminology. While many researchers continue (as I do) to use the terms "popular science" and "science popularization" while moving beyond the "traditional" model, others have suggested a change in wording. For example, in 1985 Terry Shinn and Richard Whitley argued for a new term and concept—"scientific exposition"—which, unlike the term "scientific popularization," does not carry with it the associations of the traditional model (Preface vii-viii). Their collection of analyses of popularizations is correspondingly entitled *Expository Science: Forms and Functions of Popularisation*. Literary critic Murdo McRae, editor of a later collection, favours the term "literature of science" for similar reasons ("Introduction" 10–11). The larger term "science communication" obviates the need to categorize an exchange as "popular" or not—it describes *all* communications which deal with science—and is preferred by some researchers for this reason (van Dijck 186). This is complicated, however, by the fact that "communication studies" has recently emerged as a specific discipline with its own methodologies and approaches, so the term might suggest an analytical framework that is not always intended.

My own approach is indebted to the critique of the "traditional" model I have described. My analysis views popular physics books as texts—and as producers of knowledge—in their own right, and is not concerned with comparing them with professional science writing. I do not assume, however, that the relationship between popular and professional science is of no interest at all. While it is highly limiting to restrict analyses of popular science to issues of successful or unsuccessful translation, the question of "how arguments on scientific issues travel from text to text" remains relevant in some contexts (Fahnestock, "Preserving the Figure" 6). This is true of the popularization of any specialist discipline: when literary critics or science studies researchers are criticized in the media for versions of "postmodernism" or "deconstruction" to which they do not subscribe, these issues come to the fore. Chapter 4 explores literary critics' use of popularizations in terms that both acknowledge the nature of popular physics books as producers of knowledge and maintain the importance of accuracy of information transfer in certain cases. A very useful summary of my understanding of the relationship between professional and popular science writing is provided by a literary metaphor suggested by Jon Turney. Turney argues that a "more helpful description" of the popularization process than simply "translation" is the translation of a poem, which incorporates the idea of "re-creation" ("Accounting for Explanation" 331). A translated poem, far more obviously than translated prose, produces new effects and new understandings, while maintaining a close relationship with the original text. The translated poem would not exist without the original, but either can be studied in its own right. Knowledge of the translation can affect the reading of the original, and vice versa. The degree to which the translated poem resembles the original is one question that can be asked, but not the only question.

To return to issues of definition: the texts which I here categorize as "popular physics books" are full-length expositions of physics primarily aimed at a general readership, written in the form of a monograph rather than a dictionary or encyclopedia. It should be clear from the above that these represent a *subset* of a large range of expository exchanges that might be termed "popularization." Nonetheless, the category of "full-length expositions of physics primarily aimed at a general readership" still requires qualification and explanation. As editor of popular science books Michael Rodgers points out, "We speak of a 'general readership' but what does this mean? Scientists from other disciplines? Humanities graduates? Or, as Hawking is reported to have wanted to reach, plumbers and butchers (as well as doctors, lawyers and science students)?" (232). The readership I refer to is non-specialist one—those who have no expert training in physics. This is not to imply that scientists do not read popularizations of their own discipline. Danette Paul's article "Spreading Chaos" gives a useful analysis of the way in which popularizations are taken up within the scientific community as well as outside of it. Popularizations can, she argues, have a "substantial effect" within the field they describe (61)—an example of the feedback process mentioned above. Her empirical study indicates that James Gleick's popularization *Chaos*, for instance, is used by scientists (including physicists) "both as a teaching tool and as a credible source for research" (34). Lewenstein also gives several examples of popular science books that contributed "to the intellectual development of science itself" ("How Science Books" 72–3). Popularizations can thus be deliberately deployed by scientists to make others in their own discipline, or other scientific disciplines, aware of particular ideas. This is especially useful if these ideas are too speculative or unusual to find a place in professional journals. However, scientists are rarely the *primary* readership at which a popularization is aimed; while scientists may have read *Chaos*, Gleick's popularization would not have remained on the bestseller lists for nearly a year if its primary readership had not extended well beyond the scientific community.

The question of how one determines whether a text is "aimed" at a general readership is also a problematic one. Indications of a lay readership are provided by a combination of factors: the author's discursive signals (such as colloquialisms and avoidance of jargon) as well as explicit statements about the intended readership; lack of mathematical content; and various commercial strategies, including the cover artwork, the layout of the text, its title, the stated classification of the book, and where it is placed in bookshops. However, these signals can contradict each other within a single text. In his preface to his popularization *The First Three Minutes*,

Steven Weinberg defines his reader as "one who is willing to puzzle through some detailed arguments, but who is not at home in either mathematics or physics. ... I picture the reader as a smart old attorney who does not speak *my* language, but who expects nonetheless to hear some convincing arguments before he makes up his mind" (10). Weinberg's "smart old attorney" is not representative of the majority of the "general public," and is some distance from Hawking's "plumbers and butchers." Yet the appearance of the 1983 Fontana paperback edition of Weinberg's book, with its bold front cover showing what looks like a cosmic electrical storm and sporting an endorsement from science fiction writer Brian Aldiss, emphasizes its accessibility. The blurb on the back cover claims that "even the most unscientific reader" will be able to understand the Big Bang as presented in Weinberg's "clear language." Moreover, Weinberg's book was a bestseller, and so was certainly popular in the more general sense of the word. Similarly, Roger Penrose in *The Emperor's New Mind* ignored Hawking's well-known guideline that each equation in a popular science book halves its sales (*A Brief History* vi–vii), and wrote a very dense and mathematically complex book, which became a bestseller, selling 110,000 copies in paperback in Britain between 1989 and 1996 (Sulkin). Moreover, the more superficial "signals" of a book's assumed readership such as the cover and the layout can change between editions even while the text itself remains the same.[7] These complications, however, apply to any attempt to categorize a book by genre. In practice, publishers, authors, booksellers, readers and researchers are able to recognize texts "aimed at a general readership" while acknowledging anomalous, difficult-to-categorize cases, and I have retained the term for this reason.

Another aspect of my category of "expositions of physics for a general readership" which requires comment is the term "expositions." This term is useful first because it suggests both an explanation and an act of placing something in public view; and secondly because it relates to the genre category of the "expository mode" as defined by Felicity Mellor. In an article analysing physics popularizations, Mellor identifies three overlapping categories into which popular science books can be classed: the "narratival," in which the authors tells "the story of an episode from the history of science or the life in science of one individual," as in Dava Sobel's *Longitude* (511); the "expository," which is "structured around the exposition of a particular discipline," and in which "the emphasis is on a particular subject or theme," such as Hawking's *A Brief History of Time*; and the "investigative," which "presents journalistic investigations into a topical and controversial subject of relevance to science," such as Rachel Carson's *Silent Spring* (512). Mellor notes that it is the "expository" books that are "most unambiguously identified as popularizations of science by booksellers, publishers and the authors themselves" (512). All of the texts discussed in detail in this book can be placed primarily in the "expository" category, although they draw at times on the other two modes.

---

7   A good example of the arbitrariness of categories of science writing is Paul Davies's book *The Forces of Nature*, published by Cambridge University Press. A comparison of the layout of the first edition and the second (also published by Cambridge University Press) shows that the second edition was clearly aimed at a more specialized readership than the first, although the content of both is very similar.

Some researchers make a distinction for the purpose of analysis between popularizations written by scientists themselves about their own field of expertise, and those written by journalists and professional science writers about others' fields (e.g. Shermer 489; Knudsen 1250; Lightman "'Voices of Nature'"). I have not made such a distinction, but as a literary critic I am alert to the different strategies that these two types of popularizers can use. Non-scientists are more likely to establish their authority as popularizers by aligning themselves with their readers and establishing their recognition of readers' needs and levels of understanding. Scientists are more likely to establish their authority as popularizers by including allusions—blatant or subtle—to their own standing in the scientific community. These strategies are examined in Chapters 4 and 5 respectively.

Analysing Popular Science: A Multi-Disciplinary Project

Jon Turney recently observed that "we lack an effective critical vocabulary for discussing popular science books" ("More Than Story-Telling" 47). One could equally argue that we have a superfluity of vocabularies, due to the multi-disciplinary nature of analysis of science popularization. Above I indicated the variety of terms that are used alongside or interchangeably with "popular science," and outlined the way that the assumed meaning of this term differs between researchers. Any attempt to develop a more detailed "critical vocabulary" would require a thorough-going interdisciplinarity. It is not my purpose here to attempt such an interdisciplinary taxonomy of popular physics books, but rather to bring the specific tools of the literary critic to bear on some prominent examples of this genre, in order to highlight its function as one point of exchange between Snow's "two cultures." Like any analysis of science popularization, however, this investigation must take into account existing research in a range of fields, and it is useful to give an outline of this body of work here.

Science popularization has drawn the attention of a diverse range of researchers over the last few decades. Historians of science in particular have produced a substantial amount of research dealing explicitly with the topic, as well as studies of science in the public domain and the cultural reception of scientific ideas, which often involve discussion of popularization while not necessarily using this term. Histories specifically devoted to science popularization have covered particular disciplines, theories, nations and periods, readerships, modes of delivery, institutions and societies, and individual popularizers.[8] In addition to these analyses are a host

---

8   See, for example, Cooter; Kelly; Daum; Sehgal, Sangwan and Mahanti; Burnham; Kuritz; Luey; Fyfe; Astore; Brooks; Rogers; Crosland; Wetzels; and the various publications listed in my bibliography by D. Knight; Lightman; and Bensaude-Vincent. These are just a sample of the numerous publications that have appeared over the last few decades. During the 1990s several critics observed that the history of popularization of science was comparatively under-researched (Shteir 301; Cooter and Pumfrey 237–9; Bensaude-Vincent, "In the Name of Science" 320; Lightman "'Voices of Nature'" 188). However, these observations reflect a concern with the neglect of certain kinds of approaches to

of others in related humanities disciplines, including sociology of science;[9] science communication;[10] linguistics;[11] the study of rhetoric,[12] and literary studies.[13] Many of these analyses draw on philosophy of science, and philosophers of science sometimes directly engage with specific popularizations (e.g. Norris). Scientists and popularizers themselves have also contributed to the scholarly analysis of popularizations (Eger; Stone; Wasserstein; Gould, "Fulfilling the Spandrels"). While most researchers locate themselves in one of these disciplinary areas, they are usually cognisant of relevant work in neighbouring disciplines. Collections of essays are thus often multidisciplinary, and journals such as *Public Understanding of Science*, *Science as Culture*, and *Social Studies of Science* publish analyses of popularizations from a variety of disciplinary perspectives. Interest in science popularization has been particularly healthy in the last decade or so, reflecting heightened public interest in the genre itself, as well as the influence of the young discipline of science studies, with its emphasis on the cultural embeddedness of science (outlined in Chapter 3).

Literary studies, along with the neighbouring text-focused disciplines of linguistics and rhetorical studies, have made an important and longstanding contribution to the multi-disciplinary study of science popularization. This contribution is sometimes underestimated. Earlier I noted Durant's complaint in 1995 that "for too long, literary circles have all but ignored scientific writing"; the publication two years earlier of the collection *The Literature of Science*, edited by a literary critic and dominated by contributions from literary critics, is an effective counter-example. It is true, however, that before 1990, literary analyses of popular science writing (and of scientific prose more generally) are sparser, although not entirely absent.[14] R. Allen Harris's observation in a 1991 article that there is "ample

---

popularization, and certain topics within popularization, more than a total absence of interest in the movement of scientific ideas outside of specialist spheres.

9 A prominent early example is Shinn and Whitley's collection of essays, *Expository Science*.

10 Examples include Mellor; Turney "The Word", "More than Story-Telling"; Nieman; Gregory and Miller; and Lewenstein, "Was There Really a Popular Science 'Boom'?", "How Science Books."

11 These include Jenkins; G. Fuller; Moirland; Myers, "Pragmatics"; and Calsamiglia.

12 See, for example, Lessl, "The Priestly Voice," "Science"; Charney, *The Rhetoric of Popular Science*; Fahnestock, "Accommodating Science"; Myers, "Nineteenth-Century Popularizations"; Selzer; Waddell.

13 Examples of literary analyses of popular science texts include those cited in note 17 below as well as Whitworth, "The Clothbound Universe"; various essays published in McRae, *The Literature of Science*, and in Gates and Shteir; Rousseau, "Science Books"; Shteir; Gates; Blinderman; and J. Gardner.

14 Pre-1990s literary analyses of scientific writing (in addition to those mentioned below) include Anderson; and Beer, *Darwin's Plots*. Textual analyses of science writing of various kinds from the perspective of linguistics and rhetorical studies (e.g. Myers, "Nineteenth-Century Popularizations"; Lessl, "Science", "The Priestly Voice"; Prelli; Bazerman) are easier to locate in this period. See R. Harris for a description of the field of "rhetoric of science" studies up until 1991, and for a discussion of the distinction between this and neighbouring fields. Although Harris separates the two (290), there is clearly considerable overlap between "literary" and "rhetorical" analyses of science writing.

room" for a "literary criticism of science," which is as yet "only a glint in a few scholars' eyes," is a reasonably accurate assessment of the field at that date. Harris argues that the reason for this absence is the fact that "many literary scholars are very reluctant to extend their privileged use of *literature* to include scientific texts" (290). Literary studies has always granted certain canonical non-fiction prose texts the status of literature, but Harris is right in noting that this licence has not traditionally extended to scientific writing.[15] As Terry Eagleton notes, "Nineteenth-century English literature usually includes Lamb (though not Bentham), Macaulay (but not Marx), Mill (but not Darwin or Herbert Spencer)" (1). However, these divisions are not accepted unanimously, and critics have been arguing since at least the 1960s that *certain* popular science texts at least should be treated as literature. Charles Blinderman in a 1962 article analyses T. H. Huxley's "literary style," discussing his use of devices such as alliteration and personification, and concluding that his popular writings are "a profitable study" for "the literary man" (178). In 1970 Joseph Gardner took this further, making a case for treating one of Huxley's essays as a poem. Admittedly these critics argue that the work of one prominent Victorian intellectual, not the whole genre of popular science writing, should be analysed as literature. However, broad changes within literary studies in the last few decades of the twentieth century, including an opening out of the traditional canon and a turn towards non-fiction prose genres such as diaries, autobiography and travel writing, saw science writing—both professional and popular—increasingly the subject of analysis.

Thus by 1991, when Harris was calling for a "literary criticism of science," Peter Dear in his *Literary Structure of Scientific Argument* was announcing an answer to the call: "Members of language and literature departments have begun to encompass the previously sacrosanct territory of the scientific paper ..." (1). This interest in scientific writing reflected a wider interest in the relations between literature and science. Although research in this area had been produced for decades, the 1980s saw the consolidation of a recognized field of "literature and science," with its own academic society, journal, bibliography and conferences.[16] Research into science popularization is a recognized and growing part of this field, with the latest annual bibliographies listing hundreds of entries in this area. The category, however, is widely defined to include a variety of analyses of science in popular culture (including science fiction), and literary analyses of particular popular science texts were fairly rare until the early 1990s. Since then, literary critics have paid increasing

---

15 See Myers for a detailed discussion of the way in which literary critics' focus on the body of texts termed "literature" prevented them from analysing science writing as a genre in its own right (*Writing Biology* 8–14).

16 The Society for Literature and Science (now the Society for Literature, Science and the Arts, or SLSA) was founded in 1985. It publishes a journal (originally entitled *Publications of the Society for Literature and Science* or *PSLS*; since 1993, entitled *Configurations*), issues an annual bibliography of "literature and science" criticism (now online at <http://www.litsci.org/>; before 2000 in *Configurations* and *PSLS*), and holds an annual US conference and biannual European conference.

attention to popular science books,[17] although no monograph specifically devoted to literary analysis of popular science writing has yet appeared. It is important here both to acknowledge the research that literary critics have already conducted into popular science writing, and to emphasize that, in the wake of the late twentieth-century publishing boom in popular science books, there still remains "ample room" for further analysis.

While literary criticism of popular science writing overlaps considerably with studies by rhetoricians, linguists and science communication researchers, it brings its own particular ideas and assumptions to the genre, which have their specific advantages and limitations. I want to highlight that although this investigation draws on my training in physics and on research in all of the fields mentioned above, it is methodologically centred in literary criticism. The aspects of popular physics books on which I focus—figurative language, narrative, characterization—are those which interest a literary critic, and my interest in these books is primarily in their role as mediators between the literary and scientific communities. However, just as I have drawn on a wide range of disciplines in my research for this investigation, I hope that my work in turn will be of interest to those outside (as well as within) the literary community. It is likely that the analysis of science popularization will continue to be a multi-disciplinary effort, and will contribute as much to relations between various humanities and social science disciplines as it will to relations between these disciplines and the sciences.

This book consists of six chapters that are divided into two broad groups. The first three chapters situate the genre of popular physics within historical, cultural and theoretical contexts, and provide the background material necessary for an understanding of the arguments presented in the later chapters. Chapter 1 gives a brief historical overview of popular physics books, outlining some thematic and stylistic trends, and identifying some prominent texts. The purpose of this chapter is to situate the case studies in the last three chapters within the larger genre, and also to bring some perspective to claims that the recent boom in popular physics books (and popular science books generally) is a peculiarly late twentieth-century phenomenon. The successes of works by writers such as Fred Hoyle and George Gamow in the mid-twentieth century, Arthur Eddington and James Jeans during the 1920s and 1930s, John Tyndall in the Victorian period, and popularizations of Newtonian theory in the eighteenth century, to mention only a few examples, contradict this assumption. The late twentieth-century boom, however, has its own

---

17 See, for example, Gillian Beer on the popular texts of Arthur Eddington ("Eddington") and on nineteenth-century popularizers (*Open Fields*); Andrew Angyal, and also Michael Bryson, on Loren Eiseley's writing; Jack Bushnell on *A Brief History of Time*; the essays in McRae's *Literature and Science*; Carol Gartner on Rachel Carson's *Silent Spring*; and particularly Daniel Cordle, who analyses a number of contemporary popularizers alongside postmodern novels, to illuminate a larger discussion of the relationship between literature and science. Greg Myers, a linguist who has probably produced more textual analysis of popular science writing than any other critic, also aligns his methods with those of the literary critic (*Writing Biology* 14), and has written several articles dealing with science writers' use of techniques borrowed from fiction ("Fictions for Facts"; "Fictionality").

specific characteristics and concerns, coinciding as it did with the "Science Wars" and generating at times reactions that recalled Snow's diagnosis of disciplinary relations. The boom has, as Turney notes, attracted repeated notice in the media, but "little in the way of academic analysis" ("The Word" 121). Chapter 2 thus looks at the boom itself in more detail, and attempts to understand its nature, extent and cultural meaning. I suggest that the beginnings of the boom can be located in the late 1970s in the US, and somewhat later in the UK, and that it reached its peak in the 1990s. Chapter 3 then outlines the theoretical basis of the study, examining contemporary issues surrounding the "two cultures" model and other attempts to formulate cross-disciplinary relations, and suggesting a possible role for science popularization in negotiating these issues.

The latter three chapters centre on the examination of specific popularizations. Because my approach is a literary critical one, I concentrate primarily on the textual aspects of popular science books. Like Baudouin Jurdant, who also approaches popular science books as "a literary genre," I am interested in "the text itself, as it stands before the eyes of the reader, without referring it to the didactic intention of its author" (365–6). But where Jurdant is interested in identifying the stylistic conventions of popularizations as a group, my approach in these last three chapters is rather to give close readings of individual texts. Specifically, my aim is to show how these texts can be interpreted as sites through which larger cross-disciplinary exchanges are both mediated and exacerbated. I concentrate my analysis primarily on a small core of selected texts from the 1970s, 1980s and 1990s, all of which were bestsellers and highly influential texts within the genre. These are: Gary Zukav's *The Dancing Wu Li Masters* (1979), Steven Weinberg's *The First Three Minutes* (1977), Stephen Hawking's *A Brief History of Time* (1988), James Gleick's *Chaos* (1987), and M. Mitchell Waldrop's *Complexity* (1992). A number of other related popularizations are discussed in each chapter.

Each of the three latter chapters also focuses on a particular area within physics. I have adopted as a broad thematic framework Davies's categorization of the "three ultimate frontiers of physics" as "the very small, the very large and the very complex," by which he roughly refers to cosmology, quantum physics, and chaos and complexity theories ("The New Physics" 4). Of course, not every professional physicist is busy probing these "ultimate frontiers"; many work in other well-established fields that do not require in-depth knowledge of relativistic quantum mechanics or non-equilibrium thermodynamics. However, tellingly, expositions of these more "mundane" areas, which lack the "gee-whizz" sensationalism and metaphysical allure of subjects such as cosmology, are highly unlikely to be commercially viable. It is unsurprising, then, that all of the texts I have chosen to examine align themselves usefully with Davies's three categories. Thus, Chapter 4 concentrates on quantum mechanics, Chapter 5 on cosmology and Chapter 6 on chaos and complexity theories.

Each chapter therefore deals with a particular field within physics, and in each the focus is on a particular literary device. Chapter 4 explores the use of metaphor—in particular, anthropomorphic metaphor—in popularizations of quantum theory, drawing on examples from *The Dancing Wu Li Masters* and a number of other texts. I emphasize the need for literary critics to be aware of the slippages in meaning that metaphors allow, especially when they cross disciplinary boundaries. Chapter 5 looks at narrative, specifically the way in which contemporary popularizations of cosmology, such as *A Brief History of Time* and *The First Three Minutes*, employ a mythic narrative even while emphasizing the division between scientific and mythic knowledge. This is an example of popular science overlapping with "public science"—with the public negotiation of science's status in relation to other ways of understanding the world (such as literature and the humanities). Chapter 6 is an investigation of the representation of the scientist in physics popularizations, looking at the way particular stereotypes circulate between fiction and non-fiction. The texts focused on are Gleick's *Chaos* and Waldrop's *Complexity*, both of which borrow character stereotypes and other conventions from the detective novel. This chapter most explicitly illustrates the ways in which popularizers use the tools and techniques developed by writers of fiction. Each of these case studies shows a different way in which popularizations undertake a conversation with literature, and reflects back on the issues of cross-disciplinary hostilities and harmonies discussed in the first three chapters. These last three chapters thus act as evidence that popular physics can be usefully understood, and analysed, as a point of exchange between science and literature.

CHAPTER 1

Popular Physics Books: A Brief History

The late twentieth-century popular science boom is sometimes represented as a unique, unprecedented phenomenon. Jack Harris, writing in *British Book News*, states that the boom "could be called a renaissance except that nothing remotely like it has ever happened before" (508). However, the assumption that the late twentieth-century boom is entirely unprecedented overstates the case. This chapter provides some context through which to understand the recent boom. I briefly discuss popular physics book publishing in the eighteenth and nineteenth centuries, and then examine in more detail the early-to-mid twentieth century, particularly the surge of interest in the genre following Einstein's work on the theory of relativity.[1] This wider context reveals significant parallels between the recent boom and previous periods of growth in the genre. At the same time, it is important to acknowledge the specific characteristics of popular physics book publishing during the recent boom, especially in the "glamour" areas of quantum mechanics, cosmology and chaos theory. In the second half of the chapter I point out some particularly prominent texts and identify some broad trends within these three areas. The wider cultural significance of the late twentieth-century boom, and the backlash it provoked in some parts of the literary community, are then examined in Chapter 2.

This study is literary rather than historical in focus, and does not aim to give a comprehensive history of the genre. Such a history deserves, at the least, a full-length study of its own. Roger Smith's bibliography of popular physics lists nearly nine hundred items, most of which are books. The genre is thus a large one, and also historically diverse. Bernadette Bensaude-Vincent concludes her preliminary survey of twentieth-century science popularization with the observation that this activity has shown "nothing like a linear process of development" in response to either "the specialization of science" or "a public demand"; it is "a complex and multidimensional phenomenon which has periods of expansion and relative decline" depending on "social, political and cultural contexts" ("In the Name of Science" 336). These observations are equally true of popular physics books as a subset

---

1   The first part of this chapter concentrates primarily on physics popularization in Britain; however, many of the points discussed also apply to the situation in the United States. According to John Burnham, "Americans well into the 1930s continued to depend upon English popularizers to a substantial extent" (190). David Knight, after comparing nineteenth-century popularization of science in Britain with other national cases, concludes that "[o]verall, the British experience was unique but not untypical" ("Scientists" 73). See Burnham (ch. 4) and Kuritz for general overviews of science popularization in nineteenth-century America. Daniel Kevles's history of physics in the United States includes discussions of popularization at various points in the nineteenth and twentieth centuries.

of science popularization. Bensaude-Vincent and other historians of science have produced a wealth of research examining science popularization within particular disciplines, periods, readerships and media. Here, I draw on this body of work and my own research to provide some context for the following chapters.

Popular Physics Books Before 1900

Although expositions aimed at explaining science to a wide audience have a history as long as science itself (Meadows and Hancock 1), the unstable categories of "popular science book" and "popular physics book" become more nebulous the further one moves into the past. To begin with, the "general readership" for which these books are designed changes with historical period. Until the later nineteenth century, the expense of books of all kinds put them beyond the budget of most sectors of society, and comparatively low literacy levels also limited their reach. Thus, while popular science books (in the sense of expositions for non-specialist readers) did exist in the seventeenth, eighteenth and early nineteenth centuries, their readerships were far smaller than is the case today. Another instability stems from the historical development of what we now label science. "In the eighteenth century," writes Bensaude-Vincent, "there was simply a difference of 'styles' between the scientist and the layman. In the nineteenth century it evolved into a difference in 'languages' requiring a 'translation.' In the twentieth century there was a difference of worlds" ("In the Name of Science" 329).

Until the development of various specialized, professionalized scientific languages, "natural philosophy" was accessible to any educated person. Prior to the nineteenth century, "science" as it is now conceived did not exist as a well-defined category, and the genre of the "science book" is correspondingly difficult to define (Rousseau, "Science Books" 236). Similarly, "physics" did not itself exist as a well-defined discipline until the later nineteenth century; the word "physicist" dates from the 1840s. Phenomena such as electricity and magnetism, of much interest at the turn of the nineteenth century, were considered part of chemistry at this time (Nye 4). Even during the nineteenth century, a division between books read by scientific specialists and books read by the general public is hard to sustain. Many of that century's seminal works, such as Darwin's *Origin of Species*, assumed a general readership (D. Knight, *Natural Science Books* 190). Scientific debates (such as that about the age of the Earth) ranged across different fields, and popularizers did the same.

Within these limitations, however, it is possible to identify retrospectively a long-standing tradition of popularization of those parts of science which we now categorize as physics. While science, prior to the mid-nineteenth century, was not characterized by the specialism and professionalism that it is today, the mathematical sophistication of physics always rendered it esoteric to some extent, and thus in need of exposition. Isaac Asimov suggests that Bernard le Bovier de Fontenelle was "perhaps the first person to make a reputation in science on the basis of popular science writing alone" ("Popularizing Science"). Fontenelle's *Entretiens sur la pluralité des mondes* (*Conversations on the Plurality of Worlds*) was first published in 1686, and appeared in numerous French and English editions over the next century. Fontenelle uses a literary device—a series of conversations between a natural philosopher and a beautiful marquise—to explain Copernican astronomy. A number of prominent eighteenth-century British popularizers of astronomy, including John Harris, James Ferguson and Benjamin Martin, were directly influenced by Fontenelle (Douglas 3). Newtonian physics and heliocentric astronomy were widely popularized in England in the late seventeenth and eighteenth centuries, forming part of a new market for popular science books produced by increased leisure; "literally hundreds" of popularizations of Newtonian theory were published in the eighteenth century (Rousseau, "Science Books," 211, 215). Many of these, like Fontenelle's book, found a readership in educated women. Rousseau identifies the "coffee houses" as the primary sites of science popularization, and the origin of many popular science books (207). Particularly in demand were books designed to accompany popular science lectures, and "itinerant popularizers" worked with booksellers and printers to exploit this market (208).

In the nineteenth century, the growth of the middle classes, cheaper printing techniques, the spread of literacy and the increasing specialization of science saw the development of a mass market for popularizations (Macdonald-Ross 182). The public consumption of science in museums, zoos, exhibitions and lecture halls was complemented by the private consumption of popular science books, which catered for a variety of budgets and tastes (Bensaude-Vincent, "A Genealogy" 103). David Knight observes that although periodicals were a prominent site for the popularization of science at this stage, "[b]ooks were also crucial" ("Scientists" 77). And while various fields of science such as chemistry, geology, and biology, as well as "fringe" sciences such as mesmerism and phrenology, all generated considerable public interest during the nineteenth century (Knight, "Scientists" 83), physics continued to have significant popular appeal. Prominent physicists such as Michael Faraday, Hermann von Helmholtz, James Clerk Maxwell, William Thomson (Lord Kelvin), and John Tyndall all published popular books, usually based on public lectures they had delivered.[2] As in the previous century, however, many popularizations were written by expositors who were not themselves scientists (Rousseau, "Science Books" 214; Lightman, "'Voices of Nature'"), or were positioned on the edge of the scientific community. The latter group included female popularizers such as Mary Somerville and (later in the century) Agnes Clerke. Women also continued to form a significant readership for popular science; Myers notes that purpose-written books designed for women and children enjoyed large sales throughout the nineteenth century ("Science for Women" 173).

---

2   The connection between lectures and popular books continues, to a lesser extent, in the twentieth century. Michael Whitworth notes that the "three most important popular science books" of the modernist period—Whitehead's *Science and the Modern World*, Eddington's *The Nature of the Physical World* and Jeans's *The Mysterious Universe*—were all "revisions and expansions of lectures" ("Physics" 69). Richard Feynman is a good example of a later twentieth-century popularizer whose books were based on transcripts of lectures or anecdotes. More significant in recent decades, however, are popular science books accompanying television series, such as Jacob Bronowski's *Ascent of Man* (1973), Nigel Calder's *The Key to the Universe* (1977) and Carl Sagan's *Cosmos* (1980).

Alongside the growing professionalization of science there remained a sense of public participation (Bensaude-Vincent, "A Genealogy" 105). Popular astronomy books, for example, including those by Clerke, Thomas Dick, Richard Proctor and Robert Ball, were very successful throughout the century, reflecting the continuing opportunities for amateurs to take part in and contribute to this field (D. Knight, "Scientists" 79, 80).

Bensaude-Vincent argues that "[t]he continuity between science and common sense was ... the basic postulate which underlay and even inspired most nineteenth-century popular enterprises" ("A Genealogy" 104). T. H. Huxley famously equated science with "common sense" (4: 1–23), and prominent nineteenth-century popular lectures such as Huxley's "On a Piece of Chalk" (8: 1–36), given in 1868, and Faraday's series of Royal Institution Christmas Lectures for children, *The Chemical History of a Candle* (first given in the winter of 1848–49 and published in 1861), grounded scientific knowledge in everyday objects. Popularizers of physics similarly tended to base their expositions on observable, familiar phenomena, which could be easily demonstrated during lectures: Tyndall's highly successful *Heat: A Mode of Motion* (1863) begins with a discussion of friction, demonstrated by rubbing a piece of wood against a thermopile. And although nineteenth-century physics was dominated by the development of thermodynamics, abstract philosophical implications of this subject, such as "heat death," seem to have occupied relatively little space in popularizations of this time. Stephen Brush in *The Temperature of History* observes that nineteenth-century physics popularizations such as Tyndall's made hardly any mention of the second law's disturbing implications, in contrast to early twentieth-century popularizers such as Arthur Eddington and James Jeans (61–2). Certainly, the nineteenth-century emphasis on the mundanity and accessibility of physics contrasts markedly with the early-to-mid twentieth century, in which the intensely popularized "new physics" necessitated "a radical break with common-sense views of the world" (Bensaude-Vincent, "A Genealogy" 107).

This is not to suggest, however, that nineteenth-century physics popularizations displayed no interest at all in wider philosophical, social or religious questions. Myers notes that many of the scientist-popularizers of the day "were also arguing for a particular philosophical or social position, as well as for their discipline": Tyndall for materialism; Maxwell for "a moral and spiritual alternative" to this materialism; Thomson for "a loophole in Darwinism" ("Nineteenth-Century Popularizations" 41). Many lay-popularizers such as Dick, Proctor and Clerke were writing within an explicitly Christian world-view (Lightman, "'Voices of Nature'"; Astore). There was also significant interest in psychical research among respected Victorian physicists: J. J. Thomson, Oliver Lodge, Lord Rayleigh and William Crookes were all actively involved in the Society for Psychical Research (or SPR) (Wilson 38). Lodge went on to incorporate his spiritual beliefs into physics popularizations in the early twentieth century, and another prominent physicist and SPR member, Balfour Stewart, co-wrote with fellow physicist Peter Tait a popularization entitled *The Unseen Universe or Physical Speculations on a Future State* (1875; initially published anonymously).

The success of *The Unseen Universe*, which ran through numerous editions, demonstrates a later nineteenth-century public interest in popularizations that went beyond a "common-sense" view of the world. The book can be seen as a popularization written to promote an ideological viewpoint, specifically a refutation of the materialism favoured by Tyndall and others, and an argument for the consistency of physics with the immortality of the soul (Heimann 73). Although its style is unmistakably Victorian, Stewart and Tait's book seems to have anticipated twentieth-century popularizations in a number of ways. Its title, with its connotations of mystery and the unknown, is close to early twentieth-century titles by anti-materialists, such as Jeans's *The Mysterious Universe* (1930)—one of the working titles for which was "The Shadowland of Modern Physics" (Whitworth, *Einstein's Wake* 52)—and Eddington's *Science and the Unseen World* (1929). Like these later popularizers, Stewart and Tait take the consequences of the second law of thermodynamics as one of their central concerns.[3]

Notions of physics as something enigmatic and mysterious (rather than commonsensical) were thus evident to some degree in popularizations before the turn of the twentieth century. J. W. N. Sullivan, a prominent science writer of the 1920s, observed retrospectively that the later nineteenth century had seen the emergence of popularizers who presented science as "a catalogue of marvels" ("Popular Science" 84). He traces this desire for "marvels" back to an immediate post-Darwinian popular interest in science, which he believes was "of a different kind from the leisurely interest previously shown by the cultured classes." Whereas the earlier enthusiasm represented "more genuinely an interest in science for its own sake," the post-Darwinian interest "had a different emotional basis and was merely the diversion of an interest in religious or social questions" (83). While subjects such as biology and geology had an "inevitable and immediate" association with "violent emotions," the "more exact sciences ... seem to have compromised by specialising on 'marvels.' The 'Marvels of Science' became a familiar heading, and the unsophisticated public were stunned by figures" (84). Sullivan's observation of this prevalence of "marvel-mongering" indicates that the "gee-whizz" mode identified by John Carey as the popularizer's equivalent of the sublime (*Faber Book of Science* xvi) was well established by the late nineteenth century.

Bensaude-Vincent observes that in several European countries including Britain, "An unexpected decline of popular science literature can be noticed at the turn of the century." She connects this decline with a decrease in public confidence in science as a panacea, and a tiring of "repetitive celebrations of progress" ("In the Name of Science" 321). A similar downturn occurred in the United States (Burnham 172). By the 1920s, however, science—and particularly physics—was again very much the subject of popular interest.

---

3   Their spiritualistic theme has reappeared with a paradoxically mechanistic twist in a late twentieth-century popularization, physicist Frank Tipler's *The Physics of Immortality: Modern Cosmology, God and the Resurrection of the Dead* (1994). Tipler, like Stewart and Tait, argues that each individual's thought can persist after his or her death; but where Stewart and Tait believe this information is transferred to the "unseen" universe (209), Tipler favours advanced computers (220). Tipler discusses *The Unseen Universe* in his text and claims Stewart and Tait as his "most distinguished predecessors" (352).

The Einstein Boom

The 1920s and 1930s saw a boom in popular physics books in both Britain and the United States, a publishing phenomenon which, like the late twentieth-century boom, was clearly identified by publishers and popularizers of the time (Whitworth, "Clothbound Universe" 52–7; Gregory and Miller 29). The boom was kick-started by Albert Einstein's General Theory of Relativity (1916). Numerous critics have noted the media circus that surrounded Einstein and his theory in the years immediately following Eddington's 1919 eclipse observation, which "proved" Einstein's prediction that light rays should be bent by a gravitational field to a larger degree than that predicted by Newtonian theory.[4] Friedman and Donley refer to the "publishing boom" following the observation and state that "within five years of the eclipse expedition, scientists published a number of books popularizing the new theories"—i.e., relativity and quantum physics. They cite in particular Einstein's own popularization, *Relativity: The Special and General Theory*, published in an English translation in 1920 (seven editions were released in nineteen months), and Eddington's *Space, Time and Gravitation*, published in the same year, which was among the most successful expositions of the subject on both sides of the Atlantic (17). By the time Herbert Dingle published *Relativity for All* in 1922, there were suggestions that the craze was dying down, although popularizations continued to be published (Whitworth, "Physics" 60–61), including Bertrand Russell's *ABC of Relativity* and Sullivan's *Three Men Discuss Relativity* (both 1925).

Sociologists Harry Collins and Trevor Pinch suggest that there were a number of factors, both external and internal, behind the popularity of Einstein's theory:

> [It] had something to do with the ending of the Great War and the unifying effect of science on a fractured continent. It had something to do with the dramatic circumstances and the straightforward nature of the 1919 "proof" of relativity. And it undoubtedly had something to do with the astonishing consequences of the theory for our common-sense understanding of the physical world. (*The Golem* 27)

Whereas Huxley had emphasized science as "common sense," Einstein's theory appeared to signify incomprehensibility and esotericism. Physicist Hannes Alfvén suggests that the public were "relieved" that physical reality was comprehensible only to "Einstein and a few other geniuses who were able to think in four dimensions," and that science was henceforth "something to believe in, not something which should be understood." He argues that popularizations were a factor in this change in attitude:

> Soon the bestsellers among the popular science books became those that presented scientific results as insults to common sense. The more abstruse the better! The readers liked to be shocked, and science writers had no difficulty in presenting science in a

---

4 See Collins and Pinch, *The Golem* 43–55; they argue that Eddington's results were inconclusive, and wonder if 1919 "remains a key date in the story of relativity" because "science needs decisive moments of proof to maintain its heroic image" (52).

mystical and incomprehensible way. Contrary to Bertrand Russell, science became increasingly presented as the negation of common sense. (594)

This should not imply that popularizers alone were responsible for constructing an aura of incomprehensibility around relativity: Thomas Glick claims that "popular notions of the incomprehensibility of Einstein's theories simply reflected ... what was said of them in the scientific community," and that "most of the clichés about relativity originated with anti-relativity scientists before 1919" ("Cultural Issues" 390). Nor should relativity be seen as the only factor inciting "mystical" presentations of science at the time. One of the most prominent supporters of psychic research was Lodge, whose continual promotion of the ether theory in his books and on radio long after relativity had made the concept redundant identified him with the late-Victorian old-guard rather than the new physics. Like Stewart and Tait before him, Lodge believed that the ether could explain the continuance of life after death: for him it was "the primary instrument of the Mind, the vehicle of the soul, the habitation of the Spirit. ... the living garment of God" (*Ether and Reality* 179).

It does seem likely, however, that the abstract mathematical nature of relativity theory (as well as quantum theory) was a significant factor behind the movement of popularizers of this period away from Huxley's "common sense" and towards philosophical speculation. Michael Whitworth observes that while interest in expositions which dealt with the science of relativity had subsided by the middle of the 1920s, a market had developed for popularizations that explored the "wider implications of the new physics" ("Physics" 66, 61). He cites as examples Sullivan's popularizations, such as *Aspects of Science* (1923; second series 1925) and *The Limitations of Science* (1933) (61). The books that Whitworth identifies as the "three most important popular science books of the period" (69), A. N. Whitehead's *Science and the Modern World* (1925; 1926 in Britain), Eddington's *The Nature of the Physical World* (1928) and Jeans's *The Mysterious Universe* (1930), also dealt with the philosophical implications of the new physics. Whitworth terms *Science and the Modern World* "anomalous" as a "popular" book due to the complexity of its philosophical approach, and believes its significance is "better seen in terms of its influential readership than in terms of its popular sale" ("Physics" 70). Eddington's and Jeans's books were more readily accessible, and adopted the now-familiar format in which metaphysical speculations are saved until the closing chapters. Both popularizers later published further excursions into philosophy, *The Philosophy of Physical Science* (1939) in Eddington's case, and *Physics and Philosophy* (1942) in Jeans's.

Whitworth states that although "[i]n numerical terms, [*The Nature of the Physical World*] took science publishing to new heights" ("Physics" 70), its success was "overshadowed" by that of *The Mysterious Universe* (71). He records that Jeans's popularization, published on 5 November 1930, had by the end of that year "sold 70 000 copies in the UK, and sales continued at this level into 1931." While noting Jeans's "lucid exposition," Whitworth argues that "[t]o some extent, the book may have become a self-sustaining success, with new buyers fascinated as much by its reputation as its contents. The jacket of the [1937] Pelican edition introduces it as 'the famous book which upset tradition by making Science a bestseller'" (71).

These comments suggest that *The Mysterious Universe* can be seen as a precedent for Stephen Hawking's "unique" success with *A Brief History of Time*, which sold 150,000 copies in Britain in six months (Maddox), a figure comparable to Jeans's sales. Both books' high sales were apparently due as much to their role as status-symbols as their actual content.[5]

The Mysterious Universe is particularly significant in the context of this discussion as it anticipated the late twentieth-century popular physics boom not only in terms of its widespread success but also in its rhetorical style. Jeans opens his text by unashamedly exploiting the "gee-whizz" mode. Before leaving the first page, the reader has learned that "hundreds of thousands of earths" could be easily fitted into the average star, and a "giant star" could contain "millions of millions of earths"; the total number of stars is compared to "the total number of grains of sands on all the sea-shores of the world"; and the stars are so far apart that "[i]n a scale model in which the stars are ships, the average ship will be well over a million miles from its nearest neighbour" (1). On the following page, Jeans describes the tidal forces in the sun generated by a passing star, which would have produced "a mountain of prodigious height," especially when compared to earth's "puny tides" (2). Jeans's metaphors create a sense of loneliness and grandeur: the stars are "solitary travellers" which voyage "in splendid isolation, like a ship on an empty ocean" or "[wander] blindly through space" (1). His first chapter is peppered with Conradian adjectives and adverbs suggesting the inexpressible magnitude of the phenomena he describes: "unimaginable" (1, 5), "inexorable" (12, 14), "inconceivably" (3). On the third page this rhetoric of the sublime becomes quite explicit:

> Standing on our microscopic fragment of a grain of sand, we attempt to discover the nature and purpose of the universe which surrounds our home in space and time. Our first impression is something akin to terror. We find the universe terrifying because of its vast meaningless distances, terrifying because of its inconceivably long vistas of time which dwarf human history to the twinkling of an eye, terrifying because of our extreme loneliness, and because of the material insignificance of our home in space—a millionth part of a grain of sand out of all the sea-sand in the world (3).

Jeans goes on to question the purpose of human existence in this "indifferent" universe (3), and towards the end of the chapter confronts the inevitability of heat death:

> Is this, then, all that life amounts to? To stumble, almost by mistake, into a universe which was clearly not designed for life, and which, to all appearances, is either totally indifferent or definitely hostile to it, to stay clinging on to a fragment of a grain of sand until we are frozen off, to strut our tiny hour on our tiny stage with the knowledge

---

5 There does, however, appear to be a difference in these two books' reception on an international level: *The Mysterious Universe*, which was released in the US in the same year as its British publication (Smith 305), did not appear on the bestseller list of the US magazine *Publishers Weekly* (although Jeans's earlier popularization *The Universe Around Us* [1929] remained on the list for four weeks) (Justice), while Hawking enjoyed equal success in the US and Britain.

that our aspirations are all doomed to final frustration, and that our achievements must perish with our race, leaving the universe as though we had never been? (13)

One of the striking things about Jeans's language is its similarity to that of late twentieth-century popularizations. Weinberg, in the famous conclusion of his 1977 bestseller *The First Three Minutes* (discussed in detail in Chapter 5), not only uses similar adjectives (such as "hostile" and "tiny") to describe the accidental nature of human life and its cosmic insignificance, but also makes similar use of a stage metaphor, describing humanity's fate in terms of "farce" and "tragedy." Paul Davies begins *The Last Three Minutes* (1994) with a variation on Jeans's ship/ocean metaphor—"Our solar system is a tiny island of activity in an ocean of emptiness"—and describes the Earth as "a puny object in a universe pervaded by violent forces" (4). Echoes of Jeans's rhetorical style, with its sublime metaphors, can be heard throughout the physics popularizations of the late twentieth-century boom.

Popularizing Atomic Physics

Thus far I have been discussing the 1920s and 1930s popularization wave primarily in terms of the impact of relativity, a theory which lends itself to popular discussions of large scale phenomena such as the cosmos. In general, cosmology and astronomy appear to have dominated physics popularization in the twentieth century: Smith writes that "far more" popular books and articles have been produced on these subjects than on "any other single aspect of the physical sciences" (179). The "Astronomy and Cosmology" section of Smith's bibliography (which excludes popularizations dealing specifically with relativity) is more than twice the size of the next largest section, "Atomic, Nuclear, and Particle Physics." Smith claims the latter area has "attracted much less public interest than relativity or cosmology .... Except for flurries of popularizations following the devastation of Hiroshima and Nagasaki, Japan, in 1945 and the spate of exotic particles detected in the 1950's, subatomic particles remained mostly the territory of experts until the late 1970's" (91).

Nevertheless, it is possible to find a number of early expositions of atomic physics. This subject began to enter popularizations in the mid-1920s, when quantum theory was still in its infancy (Whitworth, "Clothbound Universe" 57). There were several popular and semi-popular books dealing with early quantum theory published during the 1920s, including Sullivan's *Atoms and Electrons* and Bertrand Russell's *ABC of Atoms* (both 1923), Lodge's *Atoms and Rays* (which combined the "new physics" with the ether concept) (1924), and Jeans's *Atomicity and Quanta* (1926). In addition, many of the books discussed above, such as *The Nature of the Physical World* and *The Mysterious Universe*, deal with quantum mechanics as well as relativity. However, as Friedman and Donley observe, "quantum theory did not enter the general culture with the headlines and drama that characterized the announcements of Einstein's relativity" (128). Although atomic and subatomic physics were widely discussed in newspapers in the 1920s, with particular attention paid to their philosophical implications, it was relativity theory that was really dominating science journalism at this time (Meadows and Hancock 72). And in the 1930s, while newspaper coverage of physics was dominated by discussion of atoms and elementary particles, it was the more tangible aspects of the subject—such as particle accelerators, radioactivity and the possibility of nuclear energy—which generated most interest at this time (Meadows and Hancock 94).

In the mid-century a number of leading quantum physicists, including Max Born, Louis de Broglie, Werner Heisenberg and Niels Bohr, published popular and semi-popular discussions of their field and its interpretation.[6] Most of these texts are relatively sophisticated in tone and argument, and several, such as Heisenberg's *Physics and Philosophy*, continue the philosophical strand in physics popularization favoured by Jeans and Eddington. Rather broader in appeal are cosmologist George Gamow's attempts to explain quantum phenomena in popularizations such as *Mr Tompkins in Wonderland* (1939) and *Mr Tompkins Explores the Atom* (1944), in which an extended anthropomorphic conceit is employed in order to explain the paradoxical and nonsensical world of the new physics. Generally, however, mid-century popularizations of atomic and subatomic physics are more pragmatic in theme, in keeping with a broad shift of focus during this period.

Popular Physics in the Mid-Twentieth Century

The first popular physics boom of the twentieth century began to decline after the publication of *The Mysterious Universe* in 1930 (Whitworth, "Physics" 82). Popular science books continued to be published, but Jeans's and Eddington's style of popularization became less interesting to readers "as the decade's political emphasis shifted attention from the philosophical to the social implications" of science (Whitworth, "Physics" 83). Bensaude-Vincent notes that this period saw the development of a "specific style of popularization" which presented science "in a broad socio-historical approach with a clear social and political agenda," rather than reporting new discoveries. This "'social relations of science' movement" was developed by a number of prominent Marxist popularizers including geneticist J. B. S. Haldane, crystallographer J. D. Bernal and biologist Lancelot Hogben (Bensaude-Vincent "In the Name of Science" 328). Gregory and Miller observe a similar decrease in science popularization in the United States following the Depression, and a growing politicization of scientists (30–31).

Like the 1930s newspaper articles examined by Meadows and Hancock, many mid-century physics popularizations focus on the comparatively more practical, tangible features of the field; they give short shrift to the paradoxes and "mysteries" of quantum physics so popular later in the century. Their titles are generally based on the terms "atom" or "atomic," or take a more general form; "quantum" at this stage was clearly an esoteric term confined to physicists, academic historians and philosophers of science.[7] Unsurprisingly, the most common practical consequences

---

6 For more details, see the popularizations in my "Works Cited" listed under these authors' names.

7 Typical titles include Isaac Asimov's *Inside the Atom* (1956), Alfred Romer's *The Restless Atom* (1960), and Otto Frisch's *Meet the Atoms: A Popular Guide to Modern Physics* (1947) and

of atomic physics discussed in popularizations of the 1940s and 1950s are nuclear weapons and nuclear energy. Whereas the "new physics" had nothing whatsoever to do with the technology of the First World War, atomic and nuclear physics were closely associated in the public mind with nuclear weapons and the arms race after the Second. The opening sentence of Heisenberg's *Physics and Philosophy* is testimony to the feeling of this era: "When one speaks today of modern physics, the first thought is of atomic weapons" (27). There were a number of popularizations in the 1940s and 1950s totally devoted to the science behind the atomic bomb and nuclear energy (see, for example, Compton's *Atomic Quest* [1956]); but even popularizations which concentrate predominantly on basic concepts of atomic and subatomic physics almost inevitably culminate in chapters on atomic weapons (and to a lesser degree, atomic power). For example, the title of Selig Hecht's *Explaining the Atom* (1947) indicates a fairly general exposition of the structure of the atom, but (in the 1955 revised edition) half of the book is given over to an explanation of atomic weapons. Similarly, the final third of Joseph Feinberg's *The Atom Story* (1952) and the final quarter of Werner Braunbek's *The Drama of the Atom* (1958) are devoted to discussions of the atomic bomb. On opening Feinberg's book, the reader is immediately confronted with a glossy map showing the predicted effects on England of a hydrogen bomb dropped on London. Otto Frisch, in the introduction to his *Atomic Physics Today*, makes a point of explaining his unusual decision "after careful consideration, not to include atomic weapons" in his popularization (v), stating that he believes the subject too important to be treated peripherally (vi); discussion of atomic energy, however, takes up a quarter of his book. In contrast, late twentieth-century popularizations of subatomic physics rarely devote much space to its practical applications.

The post-Hiroshima threat to the traditionally neutral image of science was countered, according to Bensaude-Vincent, by "a huge campaign of public communication"—particularly concerning the peacetime use of atomic energy—in which science "was idealized, being depicted as a rational and context-free enterprise." In the 1950s, she claims, "popularizing essentially meant persuading the public of the social value of science" ("In the Name of Science" 331). Certainly the tone of some post-war popularizations has echoes of the optimistic clichés of the stereotypical 1940s or 1950s newsreel: "But those little pellets of atomic power [uranium] will do more than save Mr. Average Citizen his present gasoline and coal bills. For larger pellets will be used to turn the wheels of industry and when they do that they will turn the Era of Atomic Energy into the Age of Plenty" (Dietz 15).

The post-war period also saw the emergence of a number of very prominent popularizers. Eminent cosmologists Fred Hoyle and George Gamow both

---

Atomic Physics Today (1961). There are, of course, exceptions to this rule: for example, Banesh Hoffmann's *The Strange Story of the Quantum: An Account for the General Reader of the Growth of the Ideas Underlying Our Present Atomic Knowledge* (1947). In his preface, Hoffmann recommends his text to readers who want to explore the theories behind "the mysterious world of the atom" (ix). As indicated by Hoffmann's adjective "mysterious," and his use of the words "strange" and "quantum" in his title, his text is closer in theme and style to popularizations of the 1980s than to those of the 1950s, and it was in fact reissued in 1980.

authored a number of popularizations, and used their public profiles to promote their opposing views of the creation of the universe (Hoyle supported the steady state model, Gamow championed the Big Bang) (White and Gribbin 66–7; Gamow *The Creation of the Universe* 5). In the 1950s Arthur C. Clarke began writing the first of numerous popularizations of many areas of science, as did the prolific Isaac Asimov; both became high-profile science fiction writers as well as popularizers. The 1960s also saw the emergence of two prominent US-based popularizers. Nobel Prize-winning physicist Richard Feynman is famous for (among other things) his popular lectures, set down in *The Character of Physical Law* (1965), *QED* (1985) and (posthumously) *The Meaning of It All* (1998), and his two highly successful collections of anecdotes *"Surely You're Joking, Mr. Feynman!"* (1985) and *"What Do You Care What Other People Think?"* (1988). Journalist Martin Gardner, well known over the last few decades for his books and columns devoted to mathematical puzzles, produced two prominent physics popularizations in the 1960s, *Relativity for the Million* (1962) and *The Ambidextrous Universe* (1964). The influence of these writers on the present generation of popularizers is clear: Gardner wrote the foreword to Penrose's *The Emperor's New Mind*; Gribbin acknowledges the influence of Asimov and Gamow (*In Search of the Big Bang* vii, ix). Davies recalls that "the early influences on me were still from that exuberant [pre-1960s] period. Although there weren't many people writing ... they were writing in a very up-beat tone"; he mentions Hoyle in particular.[8]

More generally, however, the optimism of the 1950s was followed by "a huge disillusionment with science" due to a number of factors—the threat of nuclear annihilation, the use of chemical weapons in Vietnam and the realization that the solutions promised in the 1950s, such as the use of atomic energy, brought their own complications (Davies). This disillusionment resulted in what Davies terms "a huge turning away from interest in science popularization, and a very lean period ... the late sixties, through the seventies, even into the eighties." Lewenstein, who surveys the publication of popular science books in the United States by examining lists of prize-winning and bestselling texts, observes a roughly similar trend. He notes a slump in the genre in the three decades following World War Two, followed by a resurgence of interest in the late 1970s ("How Science Books" 72).

One of the exceptions to the slump in popular physics publishing, according to Davies, was popular astronomy; this perhaps reflected the excitement of the space-race, as well as the dissociation of this field from weapons and other immediate human concerns. In particular, black holes (the term was coined by astrophysicist John Wheeler in 1969) developed into something of a publishing and cultural phenomenon in the 1970s: "Writers picked up on the new buzz word and books appeared on the popular science and sci-fi shelves" (McEvoy and Zarate 106, 107). A science fiction film entitled *The Black Hole* was released in 1979. Hawking, one of the most prominent pioneers of the theory, "first impinged on popular awareness" at this time, featuring prominently in Nigel Calder's BBC programme *The Key to the Universe* in 1977 and its highly successful accompanying book of the same title

---

8 This and all other citations of Davies in this chapter, except where otherwise specified, refer to his responses in a personal interview, 19 January 1996.

(White and Gribbin 133, 173). John Taylor's popularization *Black Holes: The End of the Universe?* (1973) was also particularly successful; significantly, Fontana's 1975 paperback edition of this book was categorized on its back cover as "Science and Occult." This is an indication of a cultural phenomenon which, ironically, was the source of renewed interest in popular physics in the latter 1970s: the "New Age" movement.

The "New Age" of Popular Physics

The "New Age" counter-culture of the late 1960s and 1970s—a movement which typically defined itself in opposition to established science—played a prominent role in the early stages of the late twentieth-century boom in popular physics, particularly in bringing quantum mechanics to a wide readership. William Frost in *What is the New Age? Defining Third Millennium Consciousness* states that "the New Age seeks an altered state of consciousness different from the mechanistic and fragmented scientific view." This rejection of science is not wholesale, however: "Because the New Age understands itself in terms of a holistic and organic world view, the sciences are introduced as sources of the unity-aspect within the physical reality" (i). It was the perceived holistic and mystical implications of quantum mechanics, as opposed to the determinism, rationalism and reductionism of classical science, which led to its rapid adoption by New Age culture. As cultural critic Andrew Ross observes, "the larger and more humanistic sectors of the New Age community have made common cause with quantum physics, finding among the more speculative adherents of that discipline a tolerance for mysticism that complements their own holistic metaphysics, and a new *raison d'être* for closing the gap between the 'two cultures'" (*Strange Weather* 22).

As the texts by Stewart and Tait and Lodge described above indicate, mystical and paranormal ideas have a long history in physics popularizations. In 1920, Sullivan noted the appearance of "works on cosmical questions" which "seem to result from a close collaboration between, say, a professor of physics, an archdeacon and a Bond Street crystal gazer" ("The Entente Cordiale"). *The Mysterious Universe* and *The Nature of the Physical World*, while more sober than the popularizations Sullivan describes, emphasize the anti-materialist conclusions to be drawn from relativity and particularly quantum physics. Jeans famously declared that "the Great Architect of the Universe now begins to appear as a pure mathematician" and "the universe begins to look more like a great thought than like a great machine" (167, 186–7); Eddington concluded his popularization with a discussion of "Science and Mysticism." The founders of quantum physics were notable as a group for their interest in Eastern (and Western) philosophy and mysticism: Ken Wilber states that "their writings are positively loaded with references to the Vedas, the Upanishads, Taoism (Bohr made the yin-yang symbol part of his family crest), Buddhism, Pythagoras, Plato, Berkeley, Plotinus, Schopenhauer, Hegel, Kant, virtually the entire pantheon of perennial philosophers ..." (6). The physicist Lise Meitner complains in the introduction to her nephew Otto Frisch's popularization *Meet the Atoms* (1947) of the "babble of mystical conclusions and fanciful accounts" surrounding the field of atomic physics (v). However, the wider public did not appear to associate quantum theory with Eastern mysticism or paranormal notions until the emergence of New Age culture.

Signs of a specific alliance between New Age culture and quantum physics can be found in several books published in the early 1970s that present quantum mechanics as support for parapsychological speculations, such as Arthur Koestler's *The Roots of Coincidence* (1972) and psychologist Lawrence LeShan's *The Medium, The Mystic, and the Physicist: Towards a General Theory of the Paranormal* (1974). Koestler observes that in the previous few decades "theoretical physics has become more and more 'occult', cheerfully breaking practically every previously sacrosanct 'law of nature,'" and that quantum physics could be accused of "leaning towards such 'supernatural' concepts as negative mass and time flowing backwards." He concludes that "the unthinkable phenomena of ESP appear somewhat less preposterous in the light of the unthinkable propositions of physics" (11–12). LeShan's and Koestler's books clearly point to a New Age interest in quantum physics brewing in the early 1970s. It was in the later half of the decade, however, that New Age quantum physics made a significant impact upon the non-scientific public, with the publication of Fritjof Capra's *The Tao of Physics: An Exploration of the Parallels Between Modern Physics and Eastern Mysticism*.

The Tao of Physics was published originally in Britain in 1975 by Wildwood House and shortly afterwards by Shambhala in the United States (Axon 7); it became a bestseller (it has sold over one million copies worldwide, according to the cover of the 1992 Flamingo edition) and is now seen as a benchmark in science popularization. It also launched a marketing trend. In his *Beyond the Tao of Physics: Mysticism and Modern Physics—A Reappraisal*, T. J. Axon observes that "Capra's book seemed to strike a chord with the general reading public … . Publishers soon realised that there was a market for books dealing with the relationship between mysticism and modern physics and so Capra's book gave birth to a host of imitators, often works of a rather uneven quality (to say the least) …" (7). The most successful of Capra's followers was Gary Zukav, whose *Dancing Wu Li Masters* also became a bestseller; together they are typically seen as "two of the best selling gurus of pop modern physics" (Schwartz 755). Unlike the physicist Capra, Zukav is not a scientist and thus his book can be seen as a lay counterpart of *The Tao of Physics*. Neither Capra nor Zukav uses the word "quantum" in his title, and both texts also deal with relativity, but it was quantum physics that they transformed into an area of popular interest for the 1980s and 1990s. However, given the indications mentioned previously of earlier alliances between quantum physics and mysticism, it was probably the wider cultural—or counter-cultural—shift towards an embracing of mystical ideas, more than any particular insight of Capra's or Zukav's, which led to the success of their books and the subsequent boom in quantum theory popularizations.

Many reviewers and popularizers were, like Axon, concerned about the quality and reliability of popularizations which combined physics with mysticism. Physicist Jeremy Bernstein, writing in the *New Yorker*, states that *The Dancing Wu Li Masters* and *The Tao of Physics* "cannot be taken seriously as objective descriptions of modern physics," and claims, "A physicist reading these books might feel like someone on a familiar street who finds that all the old houses have suddenly turned mauve

... ." Zukav, he complains, "focusses on some of the more outré concepts in the quantum theory of measurement ... as if they were the highest achievements of modern scientific thought" ("Popular Science" 169). Raymond Sokolov in the *New York Times Book Review* assessed *The Dancing Wu Li Masters* in much the same way, remarking that "what is truly insidious about this tract-posing-as-primer is that, following the dubious notions of certain renegade physicists (who have recently come under public attack within their profession), it uses certain paradoxes of quantum theory to attack and distort mainstream physics."

It is hardly surprising then that not all, or even the majority, of the popularizers of quantum theory who followed in the wake of Capra and Zukav were keen to develop the mystical implications of their field. On the contrary, from the early 1980s many popularizers were eager to divorce themselves from this trend. Heinz Pagels in *The Cosmic Code* expresses his opinion of the New Age approach through a mini-narrative. His readers' journey on "the road to quantum reality" is interrupted by a man "with a wild look on his face": "'I've just been farther down the road, and it's really weird there. Someone told me that Einstein has proved quantum theory implies that telepathy and acausality exist—everything is connected to everything else. Quantum physicists have vindicated the ancient wisdom of the East!'" Protests that this is a distortion of Einstein's ideas, and that Buddhism is closer to classical physics than quantum theory, are lost on the original speaker, who "had already fallen asleep and wasn't listening" (147). John Gribbin states his disapproval more directly: "Quantum mechanics is identified in popular mythology, so far as it is identified at all, with the occult and ESP, some weird and esoteric branch of science that nobody understands and nobody has any practical use for." His own popularization *In Search of Schrödinger's Cat* was partly inspired, he explains, by his irritation over "the misconceptions trading under the name quantum theory among some non-scientists, Fritjof Capra's excellent *The Tao of Physics* having spawned imitators who understood neither the physics nor the Tao but suspected there was money to be made out of linking western science with eastern philosophy" (*In Search of Schrödinger's Cat* xv, xvi). Similarly, in 1984 a reviewer in the *Economist* praised Davies's *Superforce* for its avoidance of "the all-too-fashionable trap of using physics to bolster some half-baked theory that all physicists are now mystics and God is a crypto-Buddhist after all" ("Mistletoe Cutters"). "Quantum mysticism" was still drawing fire in the 1990s: physicist Leon Lederman's popularization *The God Particle* (1993) features a section entitled "The Dancing Moo-Shu Masters," in which Lederman worries that "[t]oo much of what the reading public know about physics, it knows from reading [quantum mysticism] books" (190). A systematic debunking of quantum mysticism can be found in physicist Victor Stenger's 1995 popularization *The Unconscious Quantum* (11). Stenger declares the need for "a more balanced view of the significance of developments in twentieth-century physics and cosmology" (13).

While later popularizers tried to distance themselves from "physics and mysticism" books, they were nonetheless indebted to New Age popularizers for opening the market to all kinds of popular expositions of quantum theory. In the years following the publication of Capra's and Zukav's books, quantum physics emerged from the relative obscurity it had endured for half a century. In 1984 Gribbin complained in the preface to his book *In Search of Schrödinger's Cat*,

> If all the books and articles written for the layman about relativity theory were laid end to end, they'd probably reach from here to the moon. "Everybody knows" that Einstein's theory of relativity is the greatest achievement of twentieth-century science, and everybody is wrong. But if all the books and articles written for the layman about quantum theory were laid end to end, they'd just about cover my desk. (xv)

In the mid-1980s Gribbin's observation was accurate in spirit, if a little exaggerated in actuality. It did not remain accurate for long, however; the success of Gribbin's own book was at least partly a result of a rapid reversal in the market potential of popularizations of quantum theory—mainstream as well as New Age—that was taking place as he wrote. Davies has noted the pre-1980s dearth of popularizations of quantum theory, and the change in market forces which occurred in the 1980s. He recalls that in 1980, when his popularization *Other Worlds* was released, publishers tended to view the popular physics market as limited to astronomical and cosmological subjects. He describes the difficulty he experienced when attempting to introduce quantum theory as a popular subject:

> At that particular time I didn't have an agent, I was just using a London publisher ... and they were after me writing a book on other planets or something, and I said, "I think we really want something on quantum physics." So I had to explain what that was about, and I explained about the many-universes interpretation and I said, "Well, that's sort of like other worlds," and they said, "That's great, we'll call it 'Other Worlds'... ." But I hated the title, and I said, "Look, I think we should call it 'The Quantum Factor.'" And they said, "You can't put 'quantum' in the title! People will think it's some kind of text book and it won't sell anything, no no no!"—adamantly refused, you see. And of course, about six or seven years after that, a huge number of books using the word "quantum" appeared. It's a very "sexy" word. And we really missed out there, and I've always been irritated by it.

Davies's irritation is understandable. While it is rare to find a pre-1980s popularization containing the word "quantum" in its title, the 1980s saw it develop into a marketing buzz-word, both within popular physics and elsewhere. The phrase "quantum leap," meaning "a sudden large advance," entered common usage in the 1970s (*Oxford English Dictionary*). It provided the title of a successful US television series, which initially aired between 1989 and 1993. In 1991, Bernstein noted the rapid dissemination of ideas from quantum physics into the wider culture over the previous years: "One sometimes has the feeling that much of the general public, ranging from New Age gurus to playwrights, from literary critics to futuristic economists, and on and on, have decided, as the phrase goes, that quantum mechanics is much too serious to be left to the physicists ..." (*Quantum Profiles* vii). Popularizers in a variety of fields adopted the term: Deepak Chopra in his *Quantum Healing* (1989), George Gilder in his *Microcosm: The Quantum Revolution in Economics and Technology* (1989), and Bobbi DePorter in his *Quantum Learning* (1993)—all of which pale in comparison to Kjell Enhager's *Quantum Golf: The Path to Golf Mastery* (1991) and Branton Kenton's guide to "organic gardening," *Quantum Carrot* (1987).

After the early 1980s, New Age physics books continued to flourish alongside less speculative popularizations by Gribbin, Pagels and others, and extended in a variety of directions. Michael Talbot and Fred Alan Wolf, early contributors to the "physics and mysticism" trend, continued in a similar vein. F. David Peat's *Blackfoot Physics: A Journey into the Native American Universe* (1995) looked towards connections between physics and the mythologies of indigenous peoples. Other popularizers, such as Danah Zohar, author of *The Quantum Self* (1990) and *The Quantum Society* (1993, co-authored with Ian Marshall), separated themselves from Eastern mysticism while remaining within the "general holistic movement" (*The Quantum Self* 57) characteristic of the New Age. New Age popularizations, and offshoots thereof, continued to form an integral part of the genre of quantum physics popularization throughout the popular science boom.

"Big Bang" Books: Recent Popular Cosmology

The quantum-mysticism books of the 1980s were only one component of what Davies terms a "resurgence of science popularization" at this time (the popular science boom as a cultural phenomenon is discussed in more detail in the following chapter). Cosmology continued to attract interest. Particularly prominent was astronomer Carl Sagan, whom Stephen Jay Gould termed "the greatest popularizer of the 20th century" ("Bright Star"). Sagan's book *Cosmos*, released in 1980 in conjunction with the television series of the same name, spent more than 70 weeks on the *New York Times* bestseller list (Terzian and Bilson xii).

Where quantum-mysticism popularizations presented material that had been for the most part well understood fifty years previously, trends in the popularization of cosmology tended to follow new theoretical developments or experiment results; the black hole craze mentioned above is one example. Following the discovery of the Cosmic Background Radiation in 1965, the "Big Bang" theory became well established. Weinberg's 1977 popularization of the theory, *The First Three Minutes*, became a bestseller, and was followed by several other popular books on the same topic.[9] In 1988 Hawking drew on this interest as well as the previous black hole craze (he had contributed very significantly to research in both fields) in choosing to subtitle *A Brief History of Time* with *From the Big Bang to Black Holes*. The role of this book in the late twentieth-century boom is pivotal, and the circumstances surrounding it are discussed in detail in the following chapter. Sales of *A Brief History of Time* were anomalously high compared to other popular cosmology books, but it drew on the previous ground swell of interest. Davies recalls his initial reaction to the publication of Hawking's book: "my feeling then was, 'Surely not another book on the Big Bang and all that stuff.' There had already been dozens—dozens—of books on that, some of them very good; but nobody knew about them, because the publishers didn't think there was anything in that, and they didn't bother to push them." The success of Hawking's book altered this situation; in 1991 *Publishers Weekly* suggested that the field of popular cosmology had become "a little

---

9   See, for example, Silk; Atkins; and Gribbin, *In Search of the Big Bang*.

overcrowded" (Nixon 33), and in 1995 *New Scientist* observed, "The public has an insatiable appetite for astrophysics and cosmology. Or so publishers believe—there is no shortage of such books" (R. Herbert).

During the 1990s a number of minor trends can be identified within popular cosmology. The 1991 *Publishers Weekly* article mentioned above noted a "dark matter" craze, pointing to two books published that year, Gribbin's *Blinded by the Light: The Secret Life of the Sun* (published in the US with the subtitle "New Theories About the Sun and the Search for Dark Matter") and Dennis Overbye's highly praised *Lonely Hearts of the Cosmos* (Nixon 33). The discovery in 1992 by the Cosmic Background Explorer Satellite (COBE) of irregularities in the Cosmic Background Radiation predicted by the Big Bang theory created a media sensation and launched another publishing trend, which saw the release of George Smoot and Keay Davidson's *Wrinkles in Time* (1993) and many similar popularizations.[10]

In the last few decades research into the Big Bang has brought cosmology and particle physics together, as physicists such as Hawking attempt to describe the evolution of the universe when it was very young and dense. This has resulted in the search for a Grand Unified Theory (GUT) combining three of the four fundamental forces of nature (electromagnetism and the strong and weak nuclear forces), which, in an even grander vision, would itself be combined with the fourth force, gravity, into a Theory of Everything (TOE). Hawking, at his inauguration as Lucasian Professor of Mathematics at Cambridge University in 1980, delivered a lecture entitled "Is the End in Sight for Theoretical Physics?" in which he suggested that a TOE might be possible by the end of the century (*Black Holes* 42). Smith observes that "the number of popularizations soared" to explain claims of this sort (91); they include Davies's *Superforce: The Search for a Grand Unified Theory of Nature* (1984) and Weinberg's *Dreams of a Final Theory* (1993).[11] This emphasis on discovering what is commonly termed the "Holy Grail" of physics in turn created something of a backlash: several popularizations argued that physics—or at least its "glamour" areas of particle physics and cosmology—had become overly theoretical and untestable, their titles emphasizing the "end" or "edge" of physics.[12] This backlash did not, however, quash interest in the search for a TOE, as the critical and popular success of Brian Greene's *The Elegant Universe: Superstrings, Hidden Dimensions, and the Quest for the Ultimate Theory* (1999) indicates.[13]

Davies notes that critics of the search for a TOE detected "an element of millenarianism in this heady confidence" (*Superforce* xi). Hawking's 1980 prediction that a TOE could be in reach "in the next twenty years or so" does seem conveniently well timed to tap into the public imagination. Martin Amis in his novel *London Fields* highlights this "coincidence": "Physics had died only the other day. Poor

---

10 See Rowan-Robinson; Chown; Gribbin, *In the Beginning*; and Mather and Boslough. Gregory and Miller (148–54) provide a discussion of media coverage of COBE.

11 See also Davies and Brown; Kaku and Thompson; Parker, *Search for a Supertheory*; Peat, *Superstrings*; and Barrow, *Theories of Everything*.

12 These include Morris; Boslough; Lindley; and Horgan.

13 See Turney's "Accounting for Explanation" for an analysis of Greene's book and other popularizations of superstring theory.

physics. Perhaps fifty people on earth understood it fully, but physics was over, just in time for the millennium" (296). It is hard not to interpret the spate of books expounding the search for "ultimate" or "final" explanation as a symptom of millenarian triumphalism (and similarly tempting to see the apocalyptic "End of Physics" strand as the converse response, a form of millenarian anxiety and foreboding). In his discussion of the 1990s in the collection *Fins de Siècle: How Centuries End, 1400–2000*, Asa Briggs mentions the success of *A Brief History of Time* and comments, "Time at the end of the twentieth century seemed terminal..." (224). Certainly Hawking's inaugural lecture and the title of his popularization, with its implication that humanity has reached a point in its development from which it may look back upon time itself, suggest an approaching climax of knowledge. This cultural analysis seems particularly apt given the apparent parallel concerns of the 1890s: Briggs points to a similar preoccupation with space and time during this decade, as expressed in H. G. Wells's *The Time Machine* (1895) (224), and present-day popularizers commonly observe that the late nineteenth century produced claims that physics was more or less complete. Davies, for example, writes, "By the turn of the [twentieth] century, physics had reached a curious crossroads .... In the minds of some enthusiastic physicists, the entire discipline was nearing a state of completion .... In this respect, physics at the end of the nineteenth century resembled physics at the end of the twentieth century. An all-encompassing final theory—a Theory of Everything—seemed to be within grasp" (*About Time* 14). Historians of science have questioned this view: Brush labels it "a persistent myth" and states, "Anyone who takes the trouble to read the writings of physicists and their critics in the 1890s should immediately recognize the fallacy of the myth" (*The Temperature of History* 84). Nevertheless, the fact that this myth of a false "end" to physics in the nineteenth century was promoted by popularizers at the end of the twentieth itself indicates the millenarianism surrounding the quest for a TOE.

No doubt connected to this emphasis on "ultimate explanation," as the "Holy Grail" of physics came apparently within reach, was the tendency for popularizers to delve into theological issues. As noted above, the entanglement of popular physics with religious or spiritual issues is not confined to the twentieth century, but the 1990s saw heightened interest in this area. In 1997 science writer Margaret Wertheim claimed that rhetoric associating physics—and more specifically the quest for a TOE—with God had become "endemic among physicists—as least as far as their popular writing is concerned"; her examples included Stephen Hawking, Leon Lederman, George Smoot, Paul Davies, Frank Tipler and John Polkinghorne (221). George Johnson, popularizer and science writer for the *New York Times*, stated in 1995, "Currently one of the articles of faith—or, at least, superstition—of the [popular science] publishing industry, is that one of the best ways to sell a book is to have 'God' in the title" (qtd. in Arden 7). At least two prominent physics popularizers were publishing books dealing with theology during the 1980s: Davies, whose *God and the New Physics* (1983) has since become a bestseller; and Polkinghorne, a physicist turned Anglican priest, whose excursions into science and religion include *The Way the World Is* (1983) and *One World* (1986). However, it appears that again the success of *A Brief History of Time* was the catalyst for this trend. Hawking's many casual references to God in his book, in particular his famous concluding remark that to find a TOE would be to "know the mind of God," created controversy and fuelled interest in the subject. Hawking himself noted in a later essay that if he had cut the last sentence, as he contemplated doing, "the sales might have been halved" (*Black Holes* 33). Davies suggests that Hawking never intended his remarks to have any theological seriousness, but nonetheless unintentionally "established a quasi-religious aspect to fundamental physics and cosmology." After Hawking's success Davies published another bestseller, *The Mind of God* (which takes its title from Hawking's book), and many other "God and physics" popularizations emerged. Not all of these were in earnest: Lederman's *The God Particle*, with its "Obligatory God Ending" (409), parodies more than imitates other texts in this genre. Lederman, however, has since regretted his bandwagon title, observing that it is "not a good idea" to use the word "God" in the title of a popular physics book, as it increases the difficulty of promoting rationality in the general public, and provides ammunition for the attack on objectivity from the academic community (qtd. in Arden 22). Chapter 5 looks more closely at the mythopoeic urge which lies behind physicists' tendency to present their subject to the public in a religious light.

Chaos: Popularizing a New Science

While fundamental physics and cosmology were enjoying a renaissance of sorts in the 1980s and 1990s, another important theme emerged within physics popularization: "chaos theory," and shortly afterwards, the closely related "complexity theory." It is possible to find several semi-popular expositions of chaos theory published in the 1970s and early 1980s, such as *Fractals: Form, Chance, and Dimension* (1977) and its revised version *The Fractal Geometry of Nature* (1983), by one of the early pioneers of the field, Benoit Mandelbrot. *Order Out of Chaos: Man's New Dialogue with Nature* (1984), by the Nobel laureate chemist Ilya Prigogine (another prominent researcher in the field) and science historian Isabelle Stengers, is another example.[14] However, the theory became well known to non-scientists only with the publication of Gleick's *Chaos: Making a New Science*, which became a spectacular bestseller and another benchmark in popular science publishing. Released in the United States in October 1987, *Chaos* "went through 15 printings and didn't completely disappear from national [US] bestseller lists until the following September" (Nixon 32). A number of factors may have contributed to the success of *Chaos*: Gleick's journalistic, character-based expository style; the excitement of a "new science," emphasized by Gleick's title and his explicitly Kuhnian framework; the inherent appeal of the colourful fractal images associated with chaos theory; and, perhaps, good timing. Will Nixon's observation in *Publishers Weekly* that chaos theory was a "winning topic" because it "couldn't but intrigue anyone who has noticed that we live in chaotic times" seems less banal if one remembers, as Davies does, that *Chaos* was released "the week of the stock market crash." Gleick's success turned "chaos"

---

14 See Paul's article "Spreading Chaos" for a detailed analysis of the various stages of popularization of chaos theory.

into a buzz-word: more popularizations of the theory followed, and fractals began to adorn the covers of popular physics books—even those, such as Gleick's next release *Genius: Richard Feynman and Modern Physics*, which had very little relevance to chaos theory.[15] Existing texts were remarketed to capitalize on the currency of chaos theory. *The Matter Myth: Towards 21st-Century Science*, jointly authored by Davies and Gribbin, which first appeared in 1991, was released in paperback by Penguin the following year with a fractal-adorned cover and an altered subtitle, "Beyond Chaos and Complexity." Davies's *The Cosmic Blueprint* (1987) experienced the same transformation, reappearing as a Penguin paperback in 1995 with the subtitle "Order and Complexity on the Edge of Chaos." As these subtitles indicate, "complexity" was also fast becoming a buzz-word in physics popularization. Nixon in his 1991 article notes one editor's prediction that "[c]omplexity theory, one step up from chaos, could soon inspire a shelf of books as full as that on cosmology" (32). M. Mitchell Waldrop's *Complexity: Life at the Edge of Order and Chaos*, written in a journalistic style very similar to Gleick's, appeared the next year, and was followed by a slew of other popularizations that combined chaos with complexity in their titles and subject matter.[16]

While popularizations of chaos and complexity can nominally be placed within the "popular physics" category, they are perhaps better thought of as popularizations of an area where many specialized areas of science meet. These texts are a reminder that this discussion of the history of popular physics books has focused on only one segment of a wider genre, popular science. Although Smith asserts that physics has dominated twentieth-century popular science (1), popular biology has also held a high profile for most of the twentieth century, and in its last decades genetics, neurology, evolutionary psychology and artificial intelligence were the source of numerous popular books. The 1980s and 1990s were not the decades merely of Hawking and Gleick, but also of evolutionary biologists such as Dawkins and Gould, geneticist Steve Jones, neurologist Oliver Sacks and numerous other popularizers in many areas of science. Similarly, past popularizers of other sciences, such as Lancelot Hogben and J. B. S. Haldane, are as significant as forerunners of the recent popular science boom as the physicists Eddington and Jeans. Furthermore, as I emphasized in my introduction, popular books are only a subset of a diverse range of modes and mediums for popularization, the histories of which require their own separate treatment. The texts described in the following chapters should ideally be seen in this wider contemporary and historical context: not as randomly selected textual islands, but as members of a genre, diachronically and synchronically connected to a wider discourse. This genre, however, must also be viewed in a broader cultural

---

15 The arrival of the fractal must have come as an enormous relief to editors of popular physics books. As Davies observes, the abstract concepts of the new physics are not easily represented visually, and "for a long while, almost everybody's popular science book just had a splodge of light on the front."

16 These include Lewin; Cohen and Stewart; Casti; Gell-Mann; and Coveney and Highfield.

context. In the next chapter, I address the notion of a recent popular science boom and its reception by the wider community, in particular the literary community.

CHAPTER 2

"I blame the populariscrs ...":
The Boom and Its Backlash

In my introduction I identified the recent popular science boom as one of the contexts for this investigation of popular physics books. The existence of such a boom was frequently observed by media commentators, popularizers, and publishers during the 1980s and 1990s, and is now taken as a given by researchers studying contemporary science popularization (e.g. Cassidy 127; Shermer 489; Gregory and Miller 54). However, as Jon Turney has observed, the boom itself has been the subject of very little research ("The Word" 121). The aim of this chapter is to give a rough indication of the nature and extent of the boom, and to examine the renewed "two cultures" hostilities it provoked within some sections of literary culture.

A "boom" within a particular genre (and publishing category) most obviously suggests a period of heightened activity in the genre, which in turn is usually registered by higher book sales. While there is a range of evidence that popular science book sales did increase significantly between the 1970s and the end of the century, relevant sales figures are difficult to access. In 1998 Stephen Butler, then Research Manager of the British book sales monitoring group Bookwatch, stated that the existence of the recent boom was "almost impossible to prove" quantitatively, due to such factors as the lack of data prior to the 1990s, the difficulty in categorizing and comparing books, and fluctuations in the book market as a whole.[1] Book-industry data tends not to include a specific category for popular science books. For these reasons, my discussion of sales trends in this chapter necessarily draws significantly on the observations of publishers, booksellers, journalists and popularizers, obtained in interview or in published sources.

---

1   This comment derives from a 1998 interview with Butler. Butler is one of five interviewees whose opinions I draw on in this chapter. The others are popularizer and physicist Paul Davies; Roger Highfield, Science Editor of the *Daily Telegraph;* and two publishers who have often dealt with popular science books, Will Sulkin, Publisher of Pimlico and Associate Publisher of Vintage and Jonathan Cape (all of which are imprints of Random House) and Ravi Mirchandani, an editorial director at Penguin Books. These other four interviews took place in 1996. The interviews with Davies, Sulkin and Highfield were conducted in person, in London. The interviews with Mirchandani and Butler were conducted by telephone. A telephone interview with Highfield was conducted prior to the personal interview, and I have indicated where a quotation derives from this earlier interview. More details are available in the bibliography. I have edited quotations from these interviews only to the extent of removing conversational space-fillers.

Sales, however, are only one component of a publishing boom, and do not always correlate directly with a book's cultural impact. Rex Buchanan in an article in *Publishing Research Quarterly* raises the suggestion that the impact of books "should be viewed in the far less quantifiable terms of their influence on society, or at least that component of society that makes decisions" (6). He cites as an example Rachel Carson's *Silent Spring*, which was not a bestseller when first published, but has played a significant part in the growth of the environmental movement. I believe that an important component in the recent boom is the heightened visibility and public profile of popular science books. It was this increase in the cultural cachet of these books and their authors, as much as increased sales figures, that sparked an anti-science backlash amongst members of the literary community.

What was the Popular Science Boom?

Numerous commentators writing or speaking in a diverse range of forums have observed the existence of a boom in the publishing of popular science books in the last decades of the twentieth century. Bruce Lewenstein in his article "How Science Books Drive Public Discussion" examines the history of popular science book publishing in the United States since World War Two, using the Pulitzer Prize awards and *New York Times* bestseller lists as his indicators. He notes a paucity of these books in the first thirty years after the war, followed by a significant upturn: "there is a clear change in the late 1970s. Before then, only rarely did more than 10 new science-oriented books a year become added to the list. But after 1978, only rarely do fewer than 10 science-oriented books a year get added to the list. More science books are being sold" (72). As I indicated in the last chapter, it was in the latter half of the 1970s that popular physics in particular experienced a resurgence in popularity, due in part to a perceived connection with New Age sentiments. Commentators at the time noticed the first signs of what became the boom: Jeremy Bernstein, writing in the *New Yorker* in 1979, observes "an unusual rejuvenation of interest in popular-science writing," pointing to the critical success of scientists such as Carl Sagan and Edward O. Wilson, and the interest in books such as *The Dancing Wu Li Masters* ("Popular Science" 169). By 1991, the genre was apparently so healthy that the US magazine *Publishers Weekly* could run an article entitled "The Art of Publishing Popular Science Books," asserting that "the field is flourishing" (Nixon 32).

In Britain, the boom appears to have begun somewhat later. Paul Davies identifies a "resurgence of science popularization in the eighties," but despite the enormous success in 1988 of Hawking's *A Brief History of Time*, publishers remained pessimistic about the genre's health. In 1991, one popular science editor noted:

> All of the action seems to be in the United States. ... Certainly British publishing houses have successfully published numerous popular science books of quality, but there is nothing like the same excitement here, amongst both publishers and readers, as that to be found in America. It would be nice to think that the Hawking phenomenon might begin to breathe some life into British popular science publishing and book buying, even if only slowly. (Rodgers 233–4)

This wish appears to have been swiftly granted. In a 1994 *New Scientist* article, Ravi Mirchandani, an editorial director at Penguin Books in London, claimed that the growth of popular science was "one of the most striking recent developments in general publishing"; sales figures were reaching levels that "would have seemed unimaginable" a decade earlier. In another *New Scientist* article in the same year, Alan Crowden, editorial director for science, technology and medicine at Cambridge University Press, wrote of the "current vogue" in international publishing "for books which bring science to a wider audience." When the 1996 British Association Annual Festival of Science hosted a day-long session on "Popular Science Writing in the 1990s," its accompanying leaflet began with the confident declaration, "There has been a boom in popular science publishing over the last few years ..." (Committee on the Public Understanding of Science). Alan Leitch, at that time the Marketing Director at Blackwell Retail, stated in a presentation given during the session that popular science had recently "demonstrated consistent growth" in excess of all subjects barring the Internet. Whereas in 1984 popular science was a "small niche" area, Leitch claimed, by 1994 everything had changed: "Pop science is now and continues to be big business." A 1996 article in the *Sunday Times* declared "popular-science writing" the "genre of the moment" (Porlock). The *Economist* in 1998 stated that in the previous ten years the public had been "buying science books like never before" ("Unscientific Readers" 129). Novelist A. S. Byatt, writing in the *Guardian* in mid-1998, observed that if one asks "a novelist, or a painter, or a civil servant what he or she is reading," then "the chances are that it will be a book of popular science." Around the same time, sociologist of science Steve Fuller informed *Independent on Sunday* readers that "[b]ooks popularising science have never been of a higher quality or sold better" ("The Human Touch"). Even *Vogue* readers were alerted to the phenomenon: "Forget the arts. Serious science is what's trendy now," advised a four-page 1997 article discussing Hawking, Gould, Dawkins, Steve Jones, Oliver Sacks, embryologist Lewis Wolpert, neuroscientist Susan Greenfield and former zoologist Matt Ridley (J. Turner 41).

The sales figures provided by Bookwatch for this study (see Tables 2.1 and 2.2) substantiate these claims to some extent. The tables provide data about the sales of science books in the late 1970s and the late 1990s in Britain, using the *Sunday Times* bestseller list as an indicator (they are thus similar to the figures Lewenstein uses to discuss the boom in the US). There are some difficulties in relying on these figures: year-by-year variation is significant; Bookwatch is unable to provide sales estimates for the 1970s bestsellers, hence the magnitude of the books' successes cannot be directly compared between the two lists; and the very broad definition of "science" means that the list includes several anomalous texts (one can well imagine Dawkins's reaction on finding his popularizations rubbing shoulders with *The Little Book of Feng Shui*). Keeping these qualifications in mind, it is possible nonetheless to conclude that the genre has experienced significant growth over the time period observed. A comparison of the six titles which feature in the 1978–1979 list with the twenty-six in the 1996–1999 list suggests that appearances of science books on the bestseller lists have approximately doubled. Tom Holman, who compiled the statistics, confirmed that it is "correct to identify a boom in popular science books, which have become a great deal more popular in recent times than they were

Table 2.1 Bestselling Popular Science Books, 1978–1979[a]

Title	Author	Publisher	Price (pounds)	Appearances	No.1	Highest
Manwatching	Desmond Morris	Cape	7.95	19		3
The Turin Shroud	Ian Wilson	Gollancz	5.50	12		3
Botanic Man	David Bellamy	Hamlyn	6.50	4		2
Life on Earth	David Attenborough	BBC	7.95	41	25	1
Einstein's Universe	Nigel Calder	BBC	6.25	15		2
The Voyage of Charles Darwin	Christopher Ralling	BBC	6.95	14		3

[a]The figures in the last three columns refer to the *Sunday Times* bestseller lists, and give the number of weeks the book appeared in the top ten, followed by the weeks spent at number 1, and the highest position reached.

Data compiled by Bookwatch.

Table 2.2 Bestselling Popular Science Books, 1996–1999[a]

Period	hb/pb	Title	Author	Publisher	Price (£)	Est. Sales	Appearances	No.1	Highest
96–99	pb	Men are from Mars, Women are from Venus	John Gray	Thorsons	8.99	800 000	145		2
96–99	hb/pb	Longitude	Dava Sobel	Fourth Est.	5.99	600 000	98	30	1
97–99	hb/pb	Fermat's Last Theorem	Simon Singh	Fourth Est.	5.99	170 000	30	1	1
98–99	pb	Rough Guide to the Internet	Angus Kennedy	Rgh. Guides	5.00	160 000	16	5	1
98–99	hb	The Life of Birds	David Attenborough	BBC	18.99	150 000	16	5	1
98–99	hb	Heaven's Mirror	Graham Hancock	M. Joseph	20.00	70 000	9	2	1
98	pb	The Jigsaw Man	Paul Britton	Corgi	6.99	50 000	3		9
96	pb	Fingerprints of the Gods	Graham Hancock	Mandarin	6.99	30 000	21		2
98–99	hb	The Calendar	D. Ewing Duncan	Fourth Est.	12.99	30 000	7		2
99	hb	Station X	Michael Smith	Channel 4	18.99	20 000	4	1	1
96	hb	Climbing Mount Improbable	Richard Dawkins	Viking	18.00	15 000	7		4
98	hb	Unweaving the Rainbow	Richard Dawkins	Allen Lane	20.00	15 000	1		10
98–99	hb/pb	The Mars Mystery	Graham Hancock	Penguin	6.99	15 000	4		3
99	hb	Earth Story	S. Lamb/D. Sington	BBC	19.99	15 000	1		8
98–99	pb	The Little Book of Feng Shui	Lillian Too	Element	1.99	13 000	17	1	1
98–99	hb/pb	How the Mind Works	Steven Pinker	Penguin	9.99	13 000	1		6
96	hb	Secrets of Lost Empires	M. Barnes et al	BBC	17.99	10 000	4		3
96	hb	In the Blood	Steve Jones	HarperColl.	20.00	10 000	3		7
97	hb	Stephen Hawking's Universe	David Filkin	BBC	19.99	10 000	2		3
98	hb	Legend	David Rohl	Century	20.00	10 000	1		8
98	hb	The Man Who Loved Only Numbers	Paul Hoffman	Fourth Est.	12.99	10 000	7		3
98	hb	On Giants' Shoulders	Melvyn Bragg	Hodder	12.99	10 000	3		5
99	hb	The Year 1000	R.Lacey/D. Danziger	Little, Br.	12.99	10 000	3		3
97	hb	The Human Brain: A Guided Tour	Susan Greenfield	Weidenfeld	11.99	5 000	1		10
98	hb	The Meaning of It All	Richard Feynman	Allen Lane	9.99	5 000	1		7
99	hb	Behind the Scenes at Time Team	Tim Taylor	Channel 4	18.99	5 000	2		8

[a] Sales figures provide an approximate estimate of the book's sales since publication.

Data compiled by Bookwatch.

ten or twenty years ago." As well as the number of titles on the bestseller list, it is instructive to compare the number of weeks each title spent on the list. From the tables it is clear that while five of the six 1978–1979 titles spent more than ten weeks on the lists, only seven of the twenty-six titles on the 1996–1999 list—roughly one quarter—remained on the list this long; similarly, all of the earlier titles were placed at third position or higher, while the later titles show more variation in position. The figures thus suggest that there were more moderately successful books in the late 1990s than in the 1970s. This accords with Butler's opinion that any increase in popular book sales in the last years of the century was due to the fact that more titles were currently selling reasonably well, rather than a small number of titles selling vastly more copies than previously.

By the beginning of the twenty-first century, media commentators had turned from celebrating the boom to describing its demise. In 2001, Jon Turney, at that time a science communication researcher at University College, London, stated that "the popular-science market is getting harder for publishers," arguing that the maturity of the market meant that many topics which had been new to the public in the 1970s and 1980s were now well covered, so that popularizers had to compete with their own earlier successes ("Natural Selection"). Literary agent and former publisher Peter Tallack, in an article in *Nature* published in late 2004, declares bluntly that "the phenomenon is over," noting that "the attitude of publishers and booksellers towards popular science writing has changed, and the market is not as buoyant as it was a decade ago" (803). Tallack cites market saturation and changes in booksellers' buying policies as factors in the boom's demise. He argues, however, that the genre may actually be healthier in the wake of the boom, because the popularizations that are published, while fewer in number, are higher in quality and public impact. Science popularizer Dylan Evans, writing in the *Guardian* in 2005, similarly observes that "there are signs that the popular science boom is running out of steam." He too is not unhappy about the boom's demise, as he hopes that "the flood of popular science books will go back to being the small but higher quality trickle it used to be." Thus the genre's visibility and viability may not have waned, but the boom in sales appears to have reached its peak in the 1990s.

As noted in the previous chapter, the recent boom is often described as a new, unprecedented phenomenon. Jack Harris in *British Book News* declares that "nothing remotely like it has ever happened before" (508). Several academic researchers give similar, although more qualified, appraisals. Historian Michael Shermer observes that although popular science dates back "at least to Galileo," nonetheless "In the closing decades of the 20th century, the genre … blossomed as never before" (489). Turney argues that the recent boom "goes far beyond what is frequently taken to be the previous heyday of popular writing by scientists, the 1930s, when authors like Arthur Eddington and James Jeans sold tens of thousands of copies of books about the new physics" ("The Word" 121). Such claims are endlessly debatable. In the previous chapter I noted that the early sales of Jeans's *The Mysterious Universe*—70,000 copies in 2 months—are on par with Hawking's; they are also significantly higher than the sales of popular books by established authors such as Gould or Davies (discussed below). But any attempt to quantitatively compare the success of the genre in the two periods is complicated by the limitations in sales-figure data mentioned earlier, as well as the need to take into account changes in population over the period. Moreover, the post-relativity boom is not the only previous growth period in the genre. Bernard Lightman, for example, raises the suggestion that popularization in the Victorian period could provide an appropriate historical comparison with the recent boom ("'Voices of Nature'" 187). Questions of whether the twentieth-century boom is bigger or better than previous periods of growth in the genre are impossible to prove and largely irrelevant to its analysis. Each boom period has its own specific historical circumstances and characteristics, and each is worth examining in its own right. The late twentieth-century boom could be analysed in a number of different contexts, and this study does not exhaust them. The focus here is primarily on the place of the boom, and the popular books that formed part of it, within exchanges between the literary and scientific communities.

The Role of *A Brief History*

Of all the popularizations which made up the late twentieth-century boom, none was more prominent and successful than Hawking's *A Brief History of Time*, first published in 1988. According to a 1997 *New Scientist* article, Hawking's book had by that point sold 12 million copies in English alone since its publication nine years earlier (Vines 58). It was reprinted ten times in 1988, and it remained in the *Sunday Times* bestseller list for more than four-and-a-half years (Hawking, *A Briefer History* 1). In comparison with this success, the sales of most other popular physics books are insignificant. *A Brief History of Time* was released in hardback in Britain in June 1988 (White and Gribbin 243), and by November that year had sold 150,000 copies nationally, presumably selling many more in the pre-Christmas period (Maddox). In comparison, a new paperback by Gould, one of the most prominent science popularizers of the 1980s and 1990s, would probably have sold "only between 20,000 and 30,000 in Britain in its first year" (Vines 59). According to Mirchandani, the highest-selling of Davies's books, *God and the New Physics*, had by 1996 sold a total of 70,000 copies in Britain and the Commonwealth since its publication in paperback by Penguin in 1984.[2]

Unsurprisingly, commentators unanimously consider *A Brief History of Time* a watershed in science popularization, and several suggest that due to the unique nature of its success it should be considered an anomaly within the recent history of the genre. When questioned about the book, Roger Highfield, popularizer and Science Editor of the *Daily Telegraph*, immediately repeated a joke which circulates among science writers:

Question: How many kinds of popular science books are there?
Answer: Two: *A Brief History of Time*, and all the rest. (Telephone interview)

---

2 These figures indicate that even high-selling science popularizations do not approach the success of fiction bestsellers. The two bestselling paperbacks published in Britain in the 1980s, by Sue Townsend and Jeffrey Archer, sold four hundred thousand and a million copies respectively within a year of their publication (Hamilton 262).

Will Sulkin, a publisher at Random House, stated, "There's no question, the impact of that book's phenomenal. There's also no question that it is *a* phenomenon, I mean in itself ... . No other science book has sold anything even approaching these numbers. There is no question in my own mind that this was the watershed book." In his opinion, Hawking's success was completely anomalous. Davies argued similarly:

> that whole story—Hawking, his medical condition, the book, the publishing drama and so on—is so special, in a way we should put that in a separate category. I don't think it's representative of the genre as a whole. I think the circumstances are so special that we really can't draw any general conclusions from it. ... There hasn't been another *Brief History of Time* and I very much doubt there will be.

Exactly what combination of "special" circumstances fed into Hawking's success is open to debate. This question provoked much discussion in the years following the book's publication, and continues to divide opinion, although the majority of commentators agree that Hawking's own image—a brilliant scientist coping with an extreme physical disability—was a major factor (see Chapter 5 for a discussion of this image). Michael Rodgers (an editor of popular science books at W. H. Freeman) points to three major factors which "start[ed] the ball rolling": the book's "brilliant title"; its promise to explain succinctly and non-mathematically "the origin, nature and ultimate fate of the Universe"; and the fact that "it was written by an insider of considerable status." He implies, however, that it was "'the 'Hawking effect', a public imagination captured by the thought of a brilliant mind imprisoned in a paralysed body," which was primarily responsible for its enormous success (231–2). Highfield believes there is no single easy answer to Hawking's success, which must be attributed to a variety of factors including a good title, Hawking's image, a general public fascination with cosmology (especially as told by leading scientists), the book's compact form and lack of equations, and Hawking's sense of humour (telephone interview). Sulkin favours a similar "combination of reasons," but concludes that, at bottom, the book's success is "a phenomenon which has no explanation."

One aspect of *A Brief History of Time* which most commentators believe is *not* relevant to its popularity is its readability—that is, the effectiveness of Hawking's expository style.[3] In 1991, journalist Bernard Levin announced in his column in the *Times* that "like everybody else" he was unable to go beyond page 29 of Hawking's book ("Brave Face"). It has now become a cliché that *A Brief History of Time* is the bestseller that nobody has read. Mirchandani notes the "widely discussed limitations of the book itself, and the equally widely discussed inaccessibility to many readers" ("Pop Picker"); he estimates that less than fifty percent of buyers

---

3   Exceptions include Smith, who in the introduction to his bibliography of popular physics books singles out Hawking (along with Eddington and Gamow) as one of the few science popularizers who is a "good" writer (4); and Helen Jenkins, who in a linguistic analysis of Chapter 4 of *A Brief History of Time* attempts to outline "the qualities that may have led to the book's reputation for clarity" (529), but does not provide any evidence of this reputation.

have read it through, as opposed to popularizations such as Davies's, which he estimates are read by the "vast majority" of buyers. Hawking's biographer Michael White gives a much lower figure for the readership of *A Brief History of Time*: "Farcically, it has been estimated that only 1% of buyers actually read the book." This claim is supported by the appearance of what are essentially readers' guides to the book (*Stephen Hawking for Beginners* and *Stephen Hawking's* A Brief History of Time: *A Reader's Companion*), as well as the release in 2005 of *A Briefer History of Time*, written by Hawking together with physics popularizer Leonard Mlodinow, which claims to be both a shorter and more accessible version of the original. All of these efforts would seem to defeat Hawking's original aim to write a "popular" book. As one *Times* editorial points out, it is owning—not reading—the book which is important; this might explain the contrast between the book's huge sales and its "meagre showing in the league tables of books borrowed from public libraries" ("In Time with Hawking"). *A Brief History of Time* has come to be seen by many people more as a status symbol than a text, "*the* book to be seen with in the eighties and nineties" (White and Gribbin 249).

Given the notorious unreadability of *A Brief History of Time*, it would be understandable if veteran physics popularizers such as Gribbin and Davies felt a degree of resentment towards Hawking's success. Polkinghorne acknowledges that one emotion science expositors experience in response to Hawking's triumph is "simply, envy" (*Beyond Science* 49). However, while dwarfing the sales of other popular physics books, *A Brief History of Time* simultaneously boosted the market for similar publications. White and Gribbin in their biography of Hawking assert confidently, "The popularization of science has seen a new renaissance, thanks in large measure to [Hawking's] efforts" (292). Gail Vines, a science popularizer and a consultant to *New Scientist*, states in a 1997 article, "Largely thanks to the 'Hawking effect', popular science publishing has become a serious business over the past few years" (58). A 1996 *Sunday Times* article notes "an explosion in the popularity of science books" following Hawking's success: "For a brief interlude, it seemed as though science was going to be as much a part of the Islington dinner party as sun-dried tomatoes and arguments over Bertolucci's latest film" (Connor 7). Davies believes that "publishers were so mesmerized with the runaway success of that book, that they began to desperately seek the next *Brief History of Time*."

It is widely agreed, then, that *A Brief History of Time* triggered a recognition in the publishing industry of the potential market value of popularizations, and inspired numerous attempts to repeat its success. Rodgers lists three consequences of the success of Hawking's book: a flood of more simply written copy-cat books; publishers' eagerness to make book deals with prominent scientists; and the large advances in royalties these scientists are able to obtain (233). According to Mirchandani, the increase in advances paid to popularizers in the years following Hawking's success has been even more notable than the increase in sales of popularizations. A 1990 *Nature* article, "Million-Dollar Quark," reports a record-breaking publisher's advance of one million dollars for the rights to *The Quark and the Jaguar*, a book-in-progress by Nobel laureate particle physicist Murray Gell-Mann—a deal made on the strength of a thirty-two-page proposal. The authors of the article suggest that "[r]ecovering a multi-million dollar advance may not be as difficult as it seems" (Anderson and Lincoln). Gell-Mann's literary agent, John Brockman, one of the most vocal and active supporters of the notion of a late twentieth-century renaissance within science popularization, orchestrated a number of similarly lucrative advances. According to Turney in the *Times Higher Education Supplement*, a $400,000 advance was awarded to astronomer George Smoot for his *Wrinkles in Time*, $250,000 to cosmologist Alan Guth for a proposed popularization, and $500,000 to cosmologist Frank Tipler for *The Physics of Immortality* ("New Instant Science!").

However, these advances have become notorious symbols of the over-excitement of publishers in the wake of *A Brief History of Time*. Turney describes Brockman's deals as "high-priced failures." Smoot's book, he claims, sold "under 20,000 hardback copies in the US"; Guth "could not produce the promised book"; Gell-Mann's book "went through several ghost writers and two publishers before it finally appeared, to mixed notices"; and Tipler's *Physics of Immortality* is "a monumentally silly book if ever there was one."[4] Davies believes some "very silly decisions have been made by publishers" who have "given huge advances to books of very dubious quality, [and] authors with unproven track-records, just because they're dazzled by the prospect that it could be another *Brief History of Time*."

These comments reflect the unique nature of Hawking's success, and qualify the widely held belief that *A Brief History of Time* was largely responsible for the boom's peak in the 1990s. Turney notes that "other scientists like the physicist Paul Davies and the evolutionary biologist Richard Dawkins were doing respectable business before Hawking" ("The Word" 121), and, as I outlined above, the beginnings of a boom in the United States pre-date Hawking by a decade. Moreover, it is clear that Hawking's book did not create a continuing readership for popular science in proportion to its sales success. Mirchandani argues that the huge disparity in sales between *A Brief History of Time* and later titles indicates that Hawking's readers have not continued to purchase popular science books. He questions the role of Hawking's book in launching the boom, suggesting the fact that "science has become generally trendy" as a more plausible cause:

> To publish, review or buy science books has become an important part of seeming up-to-date with what is important in current thinking .... The taste for popular science books remains a minority taste, as books are a minority taste, but it does suggest an increasing public interest in scientific ideas, and perhaps even a tendency to look as much, if not more, to scientists than, as in previous generations, political thinkers, philosophers or feminist writers, to provide a basis to their philosophical or political thinking. ("Pop Picker")

---

4   The *Times Higher Education Supplement* three weeks later published a response by Brockman, who complains that Turney's article contained a number of "factual errors." Brockman states that the figures for the advances for Tipler's and Smoot's books were incorrect, but does not supply the correct figures, which are "confidential contract matters." He notes that although Guth and Gell-Mann had their books cancelled due to inability to deliver on time, both published their books elsewhere; he insists that Gell-Mann's book was not ghost-written ("Science Writers").

Sulkin arrives at much the same conclusion via a different argument: he believes that Hawking's book did ignite the popular science boom, but that this was evident not so much in increased sales figures (a claim he questions) as in changed "public perception":

> I think that there has been a definite shift in public consciousness and awareness of science. It now [1996] occupies a more central position, rather than a peripheral position, in public consciousness than it did twenty years ago. That can be traced back absolutely directly to Stephen Hawking's book ... all that activity, and all that chatting up, as publicity departments have to do, of the media, of newspapers, and magazines, and television, raised consciousness within the media. I do think that publishing has played an instrumental role in dragging science into a fairly central position in the marketplace and in the public consciousness. Because, you know, for commercial reasons, that's what it set out to do. And it largely achieved its goal except that the result has not been that science right across the board is selling huge numbers. But nonetheless, they partially achieved their objective ... .

Likewise, for Stephen Butler of Bookwatch, the sense that popular science books had "come into their own" in the late twentieth century was due less to any obvious sales increase than a change in attitude towards these books following their promotion by media figures such as Melvyn Bragg. He believes that "mainstream publishers" in the 1990s were "looking more kindly on popular science books": they were more willing to publish them and to publicize them. These observations all point to an aspect of the popular science boom that is just as interesting as any sales increase: that is, an increase in visibility of popular science books, and a change in public perception of the value and function of these books.

The Boom and the "Two Cultures"

The notion of a changing role for popular science was promoted by two quite different commentators in the mid-1990s: John Brockman (the American literary agent responsible for the spectacular advances described above) and the British literary critic John Carey. In 1995 Brockman published *The Third Culture: Beyond the Scientific Revolution*, the result of several years of his taped conversations with scientists. Brockman terms it an "oral history of a dynamical emergent system ... an exhibition of [a] new community of intellectuals in action" (20). In his introduction he posits the existence of a "third culture," borrowing Snow's term but not his argument (17–18): whereas Snow in his 1963 lecture "The Two Cultures: A Second Look" suggested that *social* scientists might represent a link between the two cultures (Snow 69–71), Brockman's "third culture" consists of "those scientists and other thinkers in the empirical world who, through their work and expository writing, are taking the place of the traditional intellectual in rendering visible the deeper meanings of our lives, redefining who and what we are" (17). Emerging out of the "third culture," Brockman believes, is "a new natural philosophy, founded on the realisation of the import of complexity, of evolution" (20). From this philosophy comes a "new set of metaphors to describe ourselves, our minds, the universe, and all of the things we know in it," and it is the scientists producing and popularizing these "new ideas and images" who "drive our times" (21). The "third culture," Brockman argues, is not a coming together of scientific and literary intellectuals, but rather of scientists and the public; and their success, he believes, is due to the fact that "what traditionally has been called 'science' has today become 'public culture'" (18).

Simultaneously, on the other side of the Atlantic, another anthology of science writing was published: the highly successful and critically well-received *Faber Book of Science* (1995), edited by Carey. In his introduction, Carey states that "popular science has improved immensely in the later twentieth century," and that certain science popularizers, including Isaac Asimov, Freeman Dyson, Carl Sagan, Richard Feynman, Stephen Jay Gould, Stephen Hawking and Richard Dawkins, have "transformed the genre, combining expert knowledge with an urge to be understood, and bridging the intelligibility gap to delight and instruct a huge readership" (xiii–xiv). These writers, he claims, "have created a new kind of late twentieth-century literature, which demands to be recognized as a separate genre, distinct from the old literary forms, and conveying pleasures and triumphs quite different from theirs" (xiv). Although Carey states that the primary aim of the anthology is to "make science intelligible to non-scientists," he also employed aesthetic criteria in making his selections: "the first question I asked about any piece I thought of including was, Is this so well written that I want to read it twice?" (xiii). Carey believes that it is not clear "on what grounds [well-written popular science books] can be reckoned inferior to novels, poems and other representatives of the older genres" (xiv).

Although Carey's and Brockman's purposes are quite different, and their arguments diverge at many points, both emphasize the changing public perception of popular science, and implicitly position this change within the context of the "two cultures" debate. And, indeed, there is much evidence to suggest that the late twentieth-century boom owed its existence in part to a gradual opening up of the literary and artistic establishment to scientists and scientific concepts.

Several articles in *New Scientist* written in the late 1980s and early 1990s point to a "two cultures" barrier as a primary obstacle to the popularization and public understanding of science. In his 1986 article "In Praise of Science Writers," biologist Colin Tudge suggests that anyone knowledgeable about science is subject to a form of prejudice he terms "sciencism." This prejudice is evident in the media bias towards the arts:

> Thus in the heavy Sunday newspapers, which focus the proper concerns of educated people, you'll find pages and pages on the posthumous letters of Vita Sackville West, or the annotated laundry lists of D. H. Lawrence; but you'll find precious little science, past or present ... . Indeed it remains the case, just as in Snow's day, that to be interested in science is still somewhat infra dig. (44)

In his concluding paragraph Tudge argues that any "second Renaissance" of science in society would require, among other things, "a little less philistinism among the arts-trained moguls of 'the meejah'" (48). Two years later, Michael Kenward, a past editor of *New Scientist*, put forward an identical complaint:

> The disregard for books about science, not to be confused with scientific books, in newspapers and magazines is scandalous. The "literary pages" happily publish interminable reviews of the laundry lists of members of the Bloomsbury set: no massive biography (auto or otherwise) of lesser figures escapes attention, nor do the trivial confessions of media "personalities". If you are a scientist, however, the literary supplements will not even consider your book unless you have a Nobel prize or some cranky new theory about the origins of life, the Universe and everything. ("The Other Scientific Literature")

In another article written in 1990 Kenward observes, "Publishers do show some small signs of realising that science books can make money, although literary editors rarely allow reviews of them to sully their pages" ("A Prize for Plain Speaking" 68). For these commentators, the "two cultures" divide is manifested by the literary community's disregard for popular science books.

However, running parallel with these indications that the "two cultures" gap remained a relevant issue in the late twentieth century were indications of its gradual closure. As well as the numerous creative writers engaging with science (discussed in the introduction) there was growing evidence of a change in the literary establishment's attitude towards science, particularly in its popular form. In 1987, the Royal Society of Literature awarded its Heinemann Prize for the first time in its history to a popular science book, Dawkins's *The Blind Watchmaker*. In April 1999, it announced a decision to invite scientists to be fellows, as "an acknowledgment of the quality and range of contemporary science writing" (Michael Holroyd, chairman of the Royal Society of Literature, as paraphrased by Bryan Appleyard, "Mighty Minds"). The Cheltenham Literary Festival, in October of the same year, featured over a dozen science popularizers in its programme, and 2002 saw the launch of a sister event, the Cheltenham Science Festival. The interviews I conducted in 1996 also indicate a change in media attitude in the early-to-mid 1990s. Davies, while arguing that there was still "a gross imbalance" in the treatment of science and the arts within the publishing industry and the media, conceded that newspaper coverage had increased in recent years:

> Now it's a lot better than it was 10 years ago—there was a time you'd open the Sunday newspapers, you'd have ten pages of serious in-depth academic arts and literary criticism, maybe several book pages, several pages about the opera, the theatre, and so on ... and then you'd have five column inches or something on science—usually a jokey little item, with a cartoon; it was always treated as a bit of a humorous aside. To the credit of British newspapers, they have at last got themselves a half a page or something.

Mirchandani also noted an increase in the willingness of newspapers to review science books, and observed that, given the comparatively small market share of popular science, the genre received reasonable coverage. He argued that while "scientists still feel left out of what culture considers 'important,'" they "have less to complain about than, say, five years ago." Highfield agreed that although "the literary mafia is pretty tight-packed" (telephone interview), nevertheless "'steady dripping weareth away a stone'": "the Hampstead chattering classes" now acknowledged that "it's quite fun to have a few boffins strutting their stuff as well as the usual collection of poets and broadcasters and things." He suggested that the late twentieth-century boom was due more to the realization of "the arts mafia and the 'luvvie-ocracy'" of the earning potential of popular science books than with "popular science in its own right." All of these comments connect the success of popular science books in the 1990s with the lowering of perceived "two cultures" barriers.

The Backlash

The gradual lowering of these barriers, however, led some members of the literary community to raise their own. Tom Stoppard's 1993 play *Arcadia* includes an argument between a "don" of English Literature and a mathematician, who disagree over the relative significance of "personalities" and "scientific progress" (60–61). The don is clearly resentful of the encroachment of science upon the position previously held by literature, and is particularly incited by the typical subjects of physics popularizations: "Quarks, quasars—big bangs, black holes—who gives a shit? How did you people con us out of all that status? All that money? And why are you so pleased with yourselves? … I'd push the lot of you over a cliff myself. Except the one in the wheelchair, I think I'd lose the sympathy vote before people had time to think it through" (61). While Stoppard's play itself could be considered a seminal example of a text which bridges the "two cultures," his pithy portrayal of a literary intellectual's hostility towards science suggests that attitudes of this kind were still common in the 1990s.

Real-life counterparts for Stoppard's hostile literary critic are easy to locate. Two years before *Arcadia* was published, British novelist Fay Weldon wrote an article in the *Daily Telegraph* responding to a survey showing that more than ninety percent of the (British) population do not trust scientists to tell the truth. In the article she stages a debate between scientists and the public, adopting the voice of each in turn. The latter complain that scientists fail to answer the big questions: "who cares about half a second after the Big Bang; what about half a second before?" Quoting an English Literature professor's image of scientists as people with "no dress style" and "leaky pens," Weldon thinks it merciful that Snow is no longer alive to see what has happened to his "Great Divide": "It's getting really bad." She concludes with an adapted version of Louis MacNeice's "Bagpipe Music"— *"It's no go the Hawking man, / It's no go the Dawkins, / All we want is some snappy attire / And an answer or so before we expire"*—asking, "Who ever thought anything changed?" Although it is unclear from the article to what extent Weldon identifies with the view of scientists she describes, her piece generated an immediate reaction from the scientific community. Embryologist and Public Understanding of Science advocate Lewis Wolpert wrote a reply the following week, accusing Weldon of being "part of a rolling anti-science bandwagon on which so many in the humanities jump" ("So Much For Artistic License"). Lucy Ellmann's three-page article in the *Guardian* in mid-1998 (quoted at the beginning of this book) is an outright attack on scientists and particularly science popularizers. Ellmann begins by indignantly asserting that "it's not enough [for scientists] to have invented the car, the bomb, fluorescent lighting, aluminium and BSE" and to have "paved the way for the total annihilation of the planet!"; they also "want us to buy their silly popular science books" (1). Writing in a highly facetious tone, she complains of popularizers' "moustaches, terrible puns, patronising over-simplification," and the fact that "they all seem to be named *Stephen*"; their books are "[t]oo wordy, too cute, too unreliably simplistic ... ." Science, she claims, is "simply a last male stronghold, an escapist hobby for those not fully preoccupied with football"; men believe "it's a sign of virility to know your quark from your quantum" (2). As these quotations indicate, Ellmann's article is too flippant and lightweight to warrant serious rebuttal. As one angry reader observed the next week, it almost reads as a spoof (Johnson, Letter); however, the page devoted to nine hostile letters, including one from Sir Harold Kroto, suggests otherwise. The following year, journalist Brian Appleyard published an article in *Sunday Times* addressing the Royal Society of Literature's decision to invite scientists to become fellows. Appleyard accepts the decision, but warns that these fellows must be made to feel "promoted, not merely honoured." While Appleyard agrees that much science writing is "of a breathtakingly high standard," he questions whether the work of well-known popularizers such as Dawkins and Feynman has "the nuance and balance we expect of works we describe as literary" ("Mighty Minds"). In his earlier publication, *Understanding the Present* (1993), Appleyard is far more trenchant in his criticism of science popularizers. In direct contrast to Weldon, he argues that they are far too ready to address fundamental questions, questions which he believes lie outside of science's domain (see Chapter 5 for further discussion of Appleyard's argument). All of these responses are clear instances of the "two cultures" battle being fought out over the terrain of popular science.

Popularizers, unsurprisingly, argue that this hostile reaction is basically the result of envy. Davies believes that recent moves by scientists to influence opinion on "the great issues of existence" have led to "a very ugly backlash": "The fact that scientists are starting to be heard ... as evidenced by the phenomenal success of science books ... is provoking what seems to be a territorial squeal from the literary side." This reaction has "taken the form of hysterical ranting in newspapers and periodicals, and a spate of books denouncing scientists as arrogant and self-serving frauds" (qtd. in Brockman, *Third Culture* 24–5). In interview, Davies summarized the typical reaction of defensive literary intellectuals: "How dare [people like Hawking] go to number one and outsell all us worthy literary people; and how dare they do it with a book that we can't possibly understand?"[5] Cosmologist and popularizer John Barrow echoes these sentiments in an essay on science popularization, stating that "the traditional guardians of the debate about 'meaning of life' issues" are threatened by the entry of scientists into "areas that used to be their sole domain," with the result that "a minority have over-reacted against the entire scientific enterprise" ("The Analogy of Nature" 2). Several of the popularizer-scientists quoted in Brockman's book express similar opinions. Dawkins feels "somewhat paranoid" about what he sees as "a hijacking by literary people of the intellectual media" (23); in the second chapter of his own book *Unweaving the Rainbow*, he

---

5 Davies believes, however, that in the years since this 1996 interview the relationship between the literary and scientific communities in the UK has improved significantly, partly due to the efforts of Melvyn Bragg (email to the author).

mentions the same writers and journalists named by Davies, along with several others. Gould claims that there is "something of a conspiracy among literary intellectuals to think they own the intellectual landscape and the reviewing sources" (21). Psychologist Nicholas Humphrey believes, "There's terror among the British intelligentsia that culture has passed them by"; these intellectuals "learned their English literature" and looked down upon science, and now that they are "suddenly scared," they attempt to dismiss science as insignificant (25). Cosmologist Martin Rees notes, "Most of those with editorial control in the media have a primarily literary education and are now increasingly untypical, in background and interests, of intelligent readers in general" (29). Gell-Mann claims "there are people in the arts and humanities—conceivably, even some in the social sciences—who are proud of knowing very little about science and technology, or about mathematics," while "[t]he opposite phenomenon is very rare" (22).

Many of these comments echo Snow's observation that the literary establishment excludes scientists from the term "intellectual" and science from the term "culture," considering scientists "ignorant specialists" while ignoring its own disciplinary narrowness (4, 14)—an observation which itself echoes T. H. Huxley's complaint, voiced eighty years earlier in his lecture on "Science and Culture" (3: 134–59), that "the great majority of educated Englishmen" believe that "the man who has learned Latin and Greek, however little, is educated; while he who is versed in other branches of knowledge, however deeply, is a more or less respectable specialist, not admissible into the cultured caste" (141–2). Brockman emphasizes the difference between the scientists of his "third culture" and the "old-style intellectuals" they have surprised with their popular successes (18): "Unlike previous intellectual pursuits, the achievements of the third culture are not the marginal disputes of a quarrelsome mandarin class: they will affect the lives of everybody on the planet" (19). Noting Snow's observations, he argues that the exclusion of scientists from the category of "intellectual" is historically entrenched, stating that "while many eminent scientists, notably Arthur Eddington and James Jeans, also wrote books for a general audience, their works were ignored by the self-proclaimed intellectuals" (18).

Although Brockman is American, his arguments here focus on Britain; similarly, while the comments of Gould and Gell-Mann testify that this situation exists in the US, several of the scientists in the anthology believe that it is particularly relevant to Britain. Rees claims, "This problem is … even worse in the UK, because our education is more specialized …" (29). Davies believes "the problem of the two cultures" is entangled with "the class and regional prejudices that pervade British society" (24). In interview, he suggested that literary snobbishness towards science comprises the British version of the primarily US-based "Science Wars" (described in the next chapter); although the conflict manifests itself in culturally specific ways in each case, Davies sees both situations in terms of a backlash against the renewed interest in science popularizations.

At the turn of the twenty-first century, interest in these ongoing "two cultures" issues showed little sign of subsiding. In March 1999, Melvyn Bragg published a full-page lead article for the *Observer*'s "Review" section, addressing the question, "Forty years after C. P. Snow delivered his 'Two Cultures' lecture, are science and the arts any closer to ending their intellectual war?" The cartoon accompanying the article, which includes caricatures of scientist-popularizers Dawkins, Gould and Hawking (as well as Stoppard, McEwan and Frayn), and the opening paragraph, which mentions Hawking, Penrose, Gould, Dawkins, Greenfield and Steven Pinker, establish Bragg's theme. He argues that the "recent trend" towards the "media-friendly" scientist "appears only after years of a widely acknowledged rift between the arts and the sciences that has always been to the disadvantage of scientists," and he suggests that "the new love-affair" may merely be "papering over" the two cultures "chasm" (1). It may be true, he continues later in the article, that Snow is "still right" despite the fact that "Pinker, Dawkins, Hawking, Greenfield and so many others have seized the central arguments even in the most literary journals," and scientists speak on "Radio 4 talk programmes hitherto dominated by showbusiness and arts folk." Scientists "still feel that they are outside the main culture, as described by the mainstream media," and "Imagination" is ascribed "to the arts only, leaving science as the PC Plod of dull thought" (2). The following week, Radio 4 aired an hour-long debate over the motion that "This house believes that, forty years after C. P. Snow's famous lecture, Britain is still a nation of two cultures." Proposing the motion were popularizers Wolpert and Greenfield, with literary critic Gillian Beer and Simon Jenkins, author and columnist for the *Times*, opposing, and Bragg chairing. The audience decided the debate in favour of Wolpert and Greenfield by a narrow margin. In an article in the *Guardian* commenting on the debate, Tudge repeated his belief that "the gap is still there," again pointing to the arts-educated controllers of media who continue to present clichéd images of science, but also criticizing the "ostensibly populist scientists" who "have guarded their status keenly" ("Let Us Into the Tower").

Even as the popular science boom appeared to be beginning to die down, it was still generating "two cultures" concerns. In December 2002, the magazine *Prospect* (which in 1995 had featured John Carey's article "A Tale of Two Cultures" announcing popular physics as "a new literary genre" [38]) published an article entitled "The Silence of the Critics." The author, David Herman, observes that literary criticism has been replaced by (amongst other things) popular science: "On the third floor of the biggest Waterstone's in Britain is a table called 'essentials.' On two recent visits, I found it piled high with science books by Stephen Jay Gould and Susan Greenfield ... . But I found no literary criticism. It is FR Leavis's worst nightmare" (38). This article should not be considered part of an anti-science "backlash"—it argues that literary criticism has itself "gone wrong" rather than been pushed aside by science popularizers. However, it indicates ongoing concerns about what roles scientific and literary intellectuals should play in directing public culture.

What clearly emerges from all of the reactions described above, positive and negative, is the widespread belief among scientists and others that science popularization is deeply embedded in relations, both hostile and harmonious, between literature and science. Attempts to understand or analyse responses to the boom, however, often seem marred by historical short-sightedness. Ellmann states bluntly that "scientists used to be anything but popular" (2), an observation which, as my previous chapter shows, must be considered inaccurate if one's historical consciousness extends at least to the early twentieth century. Similarly, Brockman's reference to the dismissal of the popularizations of Jeans and Eddington by the literary intellectuals of their day provides an appealingly simplistic historical precedent for his arguments concerning the present. As Bensaude-Vincent notes, there were "many attempts at dialogue between science and the humanities in the inter-war period," so much so that she suggests that Snow's bemoaning of a "two cultures" divide can be read "as a kind of nostalgia of the thirties" ("In the Name of Science" 332). Literary critics have shown that the giants of literary modernism, such as T. S. Eliot, Ezra Pound and Virginia Woolf, were in contact with popularizers such as Eddington, Jeans and Sullivan and responsive to their work (Friedman and Donley 80, 96; Whitworth "Physics" 63, 121), and that science popularizations were regularly reviewed in highbrow literary journals (Whitworth "Physics" 45; 312–19). Numerous critics note the impact of relativity (and quantum physics) on the poetry and fiction of many early and mid-twentieth-century writers.[6]

There is, nonetheless, evidence that disquiet existed amongst literary intellectuals in the modernist period about scientists who were perceived to be addressing issues beyond the recognized sphere of science. Friedman and Donley note that both Eliot and Pound criticized the extrapolation of ideas from relativity into other disciplinary areas, particularly philosophy (79–80). Whitworth also records a general backlash against popularizations which attempted to combine physics and philosophy ("Physics" 74–83). In a 1920 *Athenæum* article, Sullivan notes the "unaccustomed prestige now enjoyed by science" with respect to the "impressive problems" such as "the existence of God, the 'meaning' of the Universe," and observes that, unlike "the crude materialists of Huxley's day," the modern expositor "now frequently talks to the best people, on equal terms." However, there remained among literary people, Sullivan suggests, a "nervousness" about these intruders ("The Entente Cordiale"). Thus it would seem that there is a precedent for Davies's complaint that "it's the arts and literary intellectuals who have a God-given monopoly on the great issues of existence," and that scientists who wander onto this territory are treated with suspicion and defensiveness (qtd in Brockman 24–5). However, Brockman's claim that popularizers were "ignored" by literary intellectuals in the past is certainly an overstatement.

The upshot of this discussion is that the recent popular physics boom was a real but complex phenomenon, the extent, causes and significance of which are the subject of continuing debate. What is clear, I think, is that the boom was a catalyst for—and also partly the product of—complex negotiations between representatives of the so-called "two cultures." The comments of a number of scientists and literary intellectuals mentioned above indicate that, despite the inclusive stance of writers such as Amis, Stoppard and McEwan, and the increasing openness towards science displayed by the media, some of Snow's complaints—particularly the attempted monopolization of "culture" by the literary establishment and an accompanying snobbishness towards science and its practitioners—remained relevant at the end

---

6 See, for example, Whitworth's *Einstein's Wake: Relativity, Metaphor and Modernist Literature*.

of the twentieth century, especially in Britain. In the next chapter it will become clear that this situation can be seen as a local manifestation of a wider argument between the scientific and literary communities.

CHAPTER 3

The "Two Cultures": Some Theoretical Developments

"Literary intellectuals at one pole—at the other scientists, and as the most representative, physical scientists. Between the two a gulf of mutual incomprehension …" (4). The half-century which has elapsed since Snow first proposed this bipolar model of the relationship between literature and science has seen substantial change in literary criticism. The rise of structuralism and post-structuralism, the emergence of cultural studies, and the influence of recent developments in the sociology and philosophy of science have all contributed to understandings of the nature and function of scientific discourse substantially different from those of Snow's time. Most significantly, the overturning of realist, referential views of language has significantly influenced the analysis of scientific writing, resulting in a refutation of science's traditional claim to a transparent, and hence objective, discourse. Many contemporary critics are intent on demonstrating that science is not an epistemologically privileged discourse, but is rather constructed by the wider culture. Murdo McRae, for example, asserts that "science is situated in the culture that enables it, thus science should not be exalted over literature, history, philosophy, or other nonscientific cultural expressions" (1). For these critics, the positing of science as an autonomous "culture" would implicitly deny the consequences of this realization; for with the recognition of science as a form of discourse, "it has become increasingly difficult to define precisely what science is, as opposed to, say, literature" (Levine, "One Culture" 4).

In the light of these developments, Snow's "two cultures" lecture, once the "missionary document of a new age of cross-disciplinary communication" (Bensaude-Vincent, "In the Name of Science" 332), is now seen as redundant by many critics: his concern is considered "misplaced," his model "notorious," "by now not a very helpful cliché," "an intellectual convenience—at best a paradigm whose time has passed, at worst little more than a political slogan, but at heart an inadequate image" (McRae 1; J. Black 133; Levine, "One Culture" 3; Lee 77). Current critics stress, "The distinction is one of degree, not of kind: science is no more exempt from the constraints of nonspecialist culture than literature is" (Levine, "One Culture" 3). They emphasize the extent to which criticism has moved beyond Snow's well-worn formulation—witness the titles of recent anthologies of "literature and science" criticism: *One Culture* (1987), *Beyond the Two Cultures* (1990) and *The Third Culture* (1998).

Even ignoring questions of cultural embeddedness, it is impossible to accept Snow's criteria for exchange between science and the arts. Can literary intellectuals describe the second law of thermodynamics? Are scientists familiar with the literary canon (Snow 15, 12–13)? As Levine argues, "These are not the terms of a serious debate" ("One Culture" 3). If such information-transfer represented the only obstacle to intellectual exchange, the problem could be easily remedied by force-feeding each "culture" with the corpus of Asimov or Shakespeare appropriately. Yet this is clearly not the case; and, according to science fiction novelist and literary critic Damien Broderick, bridging programmes with more sophisticated approaches have not brought about substantial change: "Despite all the subsequent attempts at makeshift interdisciplinary 'arts for engineers' and 'science for literature majors' … it is doubtful if institutional measures to remedy the problem have achieved very much of value" (4). The point at issue is not so much lists of facts belonging to each discipline that the other must commit to memory, but questions of epistemology, discursivity and methods of practice. As immunologist and Nobel laureate Peter Medawar argued in his 1968 Romanes lecture "Science and Literature," cooperation between these two fields would require their practitioners to "work their way towards an understanding—not just of each other's accomplishments … nor just of each other's purposes … but of each other's methods and energizing concepts and the quality and pattern of movement of each other's thought" (20).

Another clear objection to Snow's formulation is that it assumes intradisciplinary homogeneity. Even more so now than in Snow's time, the academic literary establishment bears a closer resemblance to a number of overlapping theoretical factions than a consensus, and the notion of a well-defined group of texts which can be termed "literature" is also highly problematic. Similarly, well before Snow's time scientific research was rapidly dividing into highly specialized disciplines and subdisciplines, a process which continues at an increasing rate. Biologist Joe Crocker, writing in *New Scientist*, argues that while a communication breakdown between science and the rest of society undoubtedly exists, "there are equally yawning gulfs between the different branches of science." What, he asks, do "physicists, geologists or engineers" know about "nematology, oncology or mycology"? Most scientists, he claims, "have trouble understanding their colleagues three doors along the corridor" (56–7).

Snow was alert to arguments of this kind. He accepted that members of the scientific culture "need not, and of course often do not, always completely understand each other; biologists more often than not will have a pretty hazy understanding of contemporary physics." In an attempt to demonstrate that intra-disciplinary unity does exist, Snow turns instead to the epistemological and methodological concerns which he tends to ignore in his criteria for interdisciplinary exchange: "there are common attitudes, common standards and patterns of behaviour, common approaches and assumptions" (9).

However, such epistemological and methodological homogeneity is not self-apparent. John Durant doubts "whether there is a single natural scientist, social scientist, historian, or philosopher who has written about the processes of scientific inquiry who would agree that these processes rest upon the twin pillars of 'the scientific attitude' and 'the scientific method'" ("What is Scientific Literacy?" 133). He cites Medawar's claim that "[t]here is indeed no such thing as the 'scientific method,'" but rather "a great variety of exploratory stratagems" which scientists employ (qtd. 133). Furthermore, as Durant suggests, these stratagems are not formally acquired, but rather learned on the job, "the same way that joiners or metal-workers learn about their respective trades" (132). Thus between specialized subdisciplinary units there is considerable room for divergence in understandings of what constitutes scientific inquiry. As Thomas Kuhn claims, science is very rarely "a single monolithic and unified enterprise that must stand or fall with any one of its paradigms as well as with all of them together," but more often appears "a rather ramshackle structure with little coherence among its various parts" (*Structure of Scientific Revolutions* 49).

On any level of analysis, then, it is problematic to assume disciplinary homogeneity within either literature or science. Moreover, disciplinary boundaries are far from stable; they are socially and historically contingent. Stefan Collini, in his introduction to a recent edition of *The Two Cultures*, observes the proliferation of interdisciplinary and subdisciplinary programmes in the last three decades, and argues, "It is largely a matter of emphasis whether one regards these changes as indicating that, rather than two cultures, there are in fact two hundred and two cultures or that there is fundamentally only one culture" (xliv). Literary critic N. Katherine Hayles argues that it is the very heterogeneity within disciplines that will enable the progress of interdisciplinary programmes such as "literature and science" criticism: "When—or if—'literature and science' becomes a discipline, let the marks distinguishing it from other departments within the academy be its awareness of fissures and discontinuities within disciplines, and its attentiveness to the complex and often conflicting assumptions that create and are created by microecologies of disciplinary niches" ("Deciphering the Rules" 45).

Hayles's approach seems to me a useful and sophisticated one; however, an emphasis on "fissures and discontinuities," if taken to extremes, would fall prey to the same flaw displayed by those who view science and literature as united in their discursive constructedness and cultural embeddedness. A reluctance to accept anomalous or liminal cases can result in the refusal to make any distinctions at all—"one culture"—or a frantic creation of more and more categories to contain disciplinary anomalies—"two hundred and two cultures." Both approaches lose in usefulness what they gain in flexibility. Snow's condemnation (in "A Second Look") of the latter view as "true" but "meaningless" (66) applies equally to the former. John Searle once observed, in a critique of Jacques Derrida's philosophy, that contemporary literary critics are, ironically, particularly prone to the positivistic assumption that "unless a distinction can be made rigorous and precise it isn't really a distinction at all." Searle argues, "On the contrary, it is a condition of the adequacy of a precise theory of an indeterminate phenomenon that it should precisely characterize that phenomenon as indeterminate; and a distinction is no less a distinction for allowing a family of related, marginal, diverging cases" (78). John Christie and Sally Shuttleworth essentially agree with this argument, remarking in their introduction to an anthology of essays on literature and science, "To reduce science to literature by insisting that science is a kind of writing, or to reduce literature to science by insisting that its codes also give a higher or privileged access to the real, are simplifications offering only the most banal of realisations. To operate across and between a primary cultural differentiation is by no means to abolish it, even wishfully." Rather, critics should "recognise the potential complexity of the terrain of literature and science once the strict and definitive boundary between them is not taken for a feature of a natural landscape, but recognised as a cultural artefact" (3).

This is no small order. Any attempt to theorize the discursive differences between literature and science must position itself within the ongoing reassessment of and debate about the relationship between scientific representation and physical reality, and thus immediately steps into the domain of philosophy and sociology of science. The nature of science as conceived by these disciplines has undergone a substantial upheaval in the decades since the 1960s. Given that these developments are central to current interdisciplinary debate and hence to this analysis, it is worth devoting some space to them here.

The Rise of Constructivism

Many critics have observed that in its early years the scientific community defined its discourse specifically in opposition to literary or figural language. According to Thomas Sprat's *History of the Royal Society* (1667), the members of the Society aimed "to separate the knowledge of Nature from the colours of Rhetorick, the devices of Fancy, or the delightful deceit of Fables" (62); they required "a close, naked, natural way of speaking; positive expressions, clear senses; a native easiness; bringing all things as near the Mathematical plainness, as they can" (113). This belief in the transparency of scientific language persisted for the next three centuries. It was integral to the "logical empiricism" (or "logical positivism") that dominated philosophy of science in the mid-twentieth century (Boyd, "Confirmation" 3, 5). As Mary Hesse describes it, the empiricist account presumed a relationship between the scientist and reality characterized by the three central assumptions of *"naïve realism,"* a *"universal scientific language"* and "the *correspondence theory of truth.*" These assumptions lead to a world-view that Hesse summarizes as follows:

> there is an external world which can in principle be exhaustively described in scientific language. The scientist, as both observer and language-user, can capture the external facts of the world in propositions that are true if they correspond to the facts and false if they do not. Science is ideally a linguistic system in which true propositions are in one-to-one relation to facts, including facts that are not directly observed because they involve hidden entities or properties, or past events or far distant events. These hidden events are described in theories, and theories can be inferred from observation, that is, the hidden explanatory mechanism of the world can be discovered from what is open to observation. Man as scientist is regarded as standing apart from the world and able to experiment and theorize about it objectively and dispassionately. (*Revolutions and Reconstructions* vii)

The parallels between this view and naïve realist theories of literature are clear; and like realist theories of literature, this approach could not withstand the interdisciplinary reassessment of language and representation that characterized the 1960s and 1970s.

Within the philosophy of science, the work of such theorists as Kuhn, Norwood Russell Hanson and Paul Feyerabend established that scientific observation is theory-laden: that "(1) observations are theory-impregnated in the sense that they involve auxiliary assumptions in the form of measurement theories, theories of psychology, theories of linguistic classification, etc.; and (2) observations are theory-impregnated in the sense that what counts as relevant and proper evidence is partly determined by the theoretical paradigm which the evidence is supposed to test" (Knorr-Cetina and Mulkay 4). This argument is analogous to that often aimed at literary critics who assert the benefits of a traditional untheorized approach to literature: "without some kind of theory, however unreflective and implicit, we would not know what a 'literary work' was in the first place, or how we were to read it" (Eagleton viii). More broadly, the "theory-ladenness" of knowledge parallels the structuralist claim that an individual's use of language is always constrained by the pre-existing categories that structure the language s/he employs. There is no innocent eye, just as there is no "naked, natural way of speaking."

Theory-ladenness does not in itself appear to warrant a startling change in scientists' understanding of their activity. Medawar argued a very similar point in his well-known 1963 radio broadcast "Is the Scientific Paper a Fraud?":

> naïve observation, innocent observation, is a mere philosophic fiction. There is no such thing as unprejudiced observation. ... all scientific work of an experimental character starts with some expectation about the outcome of the enquiry. This expectation one starts with, this hypothesis one formulates, provides the initiative and incentive for the enquiry and governs its actual form. It is in the light of this expectation that some observations are held relevant and others not; that some methods are chosen, others discarded; that some experiments are done rather than others. It is only in light of this prior expectation that the activities the scientist reports in his scientific papers have any meaning at all. (231)

However, the notion of theory-ladenness has led to developments within the philosophy, history and sociology of science that would almost certainly be unacceptable to working scientists such as Medawar. According to Richard Boyd, within philosophy of science it is now "universally accepted" that "many or all of the central methods of science are theory-dependent" ("Confirmation" 7–8), and this has led to the development of two alternatives to logical empiricism, "scientific realism" and "social constructivism" (11). Boyd explains that these two positions can be seen roughly as two alternative responses to the question, "What must the world be like in order that a methodology so theory-dependent as ours could constitute a way of finding out what's true?" The scientific realist asserts the world is "one in which the laws and theories embodied in our actual theoretical tradition are approximately true" ("On the Current Status of Scientific Realism" 207); the constructivist, however, argues that "the world which scientists study must be, in some robust sense, defined or constituted by, or 'constructed' from, the theoretical tradition in which the scientific community in question works" (202). Of these two alternatives, it is the constructivist perspective that most closely reflects developments within the humanities and social sciences, especially sociology.

While the more traditional field of sociology of science addressed itself to social relationships between *scientists*, the sociology of scientific knowledge (SSK) which emerged in the 1970s set out to analyse the social processes behind the production and acceptance of the *actual content* of scientific knowledge (Woolgar,

Science 26). SSK is centrally supported by the notion of theory-ladenness described above, as well as the concept of "underdetermination" of scientific theories by empirical evidence. According to Karin Knorr-Cetina and Michael Mulkay, this thesis, which follows from the work of philosopher W. V. Quine and others, asserts that "any theory can be maintained in face of any evidence, provided that we make sufficiently radical adjustments elsewhere in our beliefs." No theory, it is argued, can be isolated from the complex network of supporting assumptions and "auxiliary hypotheses." If observational evidence fails to support a theory, it can nevertheless be retained by adjusting these auxiliary hypotheses. The converse also holds: "there are in principle always alternative theories which are equally consistent with the evidence and which might reasonably be adopted by scientists" (3).

Together, the thesis of underdetermination of scientific theories by the data and the thesis of theory-ladenness support the argument that sociological investigation into scientific knowledge production (that is, SSK) is justified (Knorr-Cetina and Mulkay 5). Further, the last few decades have seen the emergence of a group of scholars who undertake what Steve Woolgar terms "'the social study of science' (SSS)" (*Science* 14). According to Woolgar, this new disciplinary group works on the premise that "there is no essential difference between science and other forms of knowledge production" (12). SSS is highly influenced by SSK (14), adopting the latter's emphasis on "the relativity of scientific truth" and its call for "a sociological analysis of technical content" (41). Woolgar states that SSS represents a multidisciplinary grouping, "notably sociology and history of science, less prominently, philosophy, anthropology and psychology" (14). More commonly used at present are the alternative terms "Science and Technology Studies" (STS) or simply "science studies," which are generally understood to cover a wide range of disciplines, including the sectors of cultural studies and "literature and science" criticism which share similar assumptions and aims.

As science studies describes an internally complex grouping of interdisciplinary research programmes representing a variety of related theoretical stances, any attempt to summarize its methods and goals inevitably imposes a false disciplinary homogeneity. It is important here, however, to describe some of the more prominent research programmes within science studies, since its terms and methods have been highly influential on "literature and science" criticism, yet are now often used in such a loose and casual sense that it is difficult to gain an in-depth understanding of their meaning.

I have already mentioned the close parallels between the philosophical theory of "social constructivism" and developments within the sociology of science. Specifically, within sociology the "constructivist programme" was originally formulated by Knorr-Cetina as an approach to the ethnography of scientific work, which

> considers the products of science as first and foremost the result of a process of (reflexive) fabrication. Accordingly, the study of scientific knowledge is primarily seen to involve an investigation of how scientific objects are produced in the laboratory rather than a study of how facts are preserved in scientific statements about nature. ("The Ethnographic Study of Scientific Work" 118–19)

In Knorr-Cetina's formulation, the constructivist programme involves "a *genetic* and *microlevel* approach to the problem of the social conditioning of scientific knowledge" (116–17); that is, it involves ethnographers spending time in laboratories examining the practices of scientists and understanding in detail the way in which they construct scientific knowledge. However, even within sociological circles the term "constructivist" has taken on a broader meaning. For example, while Knorr-Cetina denies that constructivism requires a "relativist position in the common ... sense of the word" (136), constructivism is nevertheless usually considered a relativistic approach. Stephen Cole observes that "the approach to the sociology of science dominant today was first called 'relativism-constructivism' and is now commonly referred to as 'social constructivism'" (4–5). Within a multi-disciplinary context, it is near-impossible to define constructivism. Siegfried Schmidt notes that constructivism is "not a homogeneous theory," and was not developed by "a homogeneous community of scholars, who agree on a common terminology"; rather, "what is labeled 'constructivism' in philosophy and in various fields of research (ranging from biology and family therapy to literary studies) is a complex discourse created by many voices coming from various sources" (295). For my purposes here, perhaps the most broad, workable definition of constructivism is that provided in *The Philosophy of Science*: "The view that the subject matter of scientific research is wholly or partly constructed by the background theoretical assumptions of the scientific community and thus is not, as realists claim, largely independent of our thoughts and theoretical commitments" (Boyd, Gasper, and Trout 775).

One social constructivist approach that has been highly influential is the "strong programme" within the sociology of scientific knowledge. The strong programme, as originally formulated by David Bloor in 1976, asserts that the aim of SSK is "to discern which conditions bring about beliefs or states of knowledge," and to explain these beliefs in the same way regardless of whether they are perceived as true or false (Woolgar, *Science* 42–3). Strong programme supporters seek the same causes for the production and acceptance of the knowledge-claims of "pseudosciences," or of research considered to be erroneous, as they do for well-established scientific theories. The strong programme thus appears to dissolve the epistemological distinction between science and other knowledge systems, and as such supports a form of relativism. It is constructivist in that it assumes that factors other than physical phenomena can account for the construction of scientific knowledge; this is implicit in its insistence on finding the same causes behind discredited and well-accepted knowledge.

It is understandable, then, that the theoretical perspective dominant in science studies should be termed "relativist-constructivist." However, it is important to discriminate between the various forms of relativism employed within science studies. Knorr-Cetina and Mulkay distinguish between "epistemic" and "judgemental" relativism. Epistemic relativism "asserts that knowledge is rooted in a particular time and culture" and does not simply "mimic nature"; thus "insofar as scientific realism wishes to make such a claim, epistemic relativism is anti-realist." By contrast, judgemental relativism "appears to make the additional claims that all forms of knowledge are 'equally valid', and that we cannot compare different forms of knowledge and discriminate among them" (5). Knorr-Cetina and Mulkay emphasize that "judgemental relativism manifestly does not follow from epistemic relativism":

> The belief that scientific knowledge does not merely replicate nature in no way commits the epistemic relativist to the view that therefore all forms of knowledge will be equally successful in solving a practical problem, equally adequate in explaining a puzzling phenomenon, or, in general, equally acceptable to all participants. (6)

They suggest that few critics argue the case for judgemental relativism (14 n. 6).

Knorr-Cetina's version of constructivism seems to leave some space for the role of the physical world in the production of scientific knowledge, even if it is not (in her words) "first and foremost" in this process. Those with a more critical approach to constructivism, such as Cole, who terms himself a "realist-constructivist," place much emphasis on the importance of leaving this space: "I do not believe that evidence from the external world *determines* the content of science, but I also reject the position that it has no influence" (x). Cole's argument against relativist-constructivists echoes Searle's criticism of literary theorists: they favour a dichotomy—science as either totally objective or totally unaffected by physical constraints—over a continuum. If we return to the philosophical definition of constructivism cited above, the issue in question here would seem to be contained in the phrase "wholly or partly constructed." The difference between "wholly" and "partly" is of paramount importance: while the view that social processes are *involved* in the production of scientific knowledge is hardly a radical one, the belief that they *totally determine* this knowledge—i.e., that the natural world places *no constraints* on scientific theories—is highly controversial. According to Cole, critics who deny the constraints of the physical world upon scientific knowledge misinterpret the theory of underdetermination. Even granting Kuhn's claim that "it is impossible to use empirical evidence to choose between two specific competitors in a paradigm debate," this "does *not* mean that it would be impossible to use such evidence to choose among any competitors. ... Because science is underdetermined does not mean that it is completely undetermined" (Cole 23, 24).

While many supporters of social constructivism do not consider scientific knowledge as wholly the product of social factors, others do appear to recommend this extreme relativist perspective. Woolgar observes admonishingly that many sociologists, including advocates of the "strong programme," are uncertain about "taking issue" with the assumption that "the world exists independently of, and prior to, knowledge produced about it" (*Science* 53). For him, Harry Collins's view that "the natural world has a *small or non-existent* role in the construction of scientific knowledge" does not go far enough (qtd. in Woolgar 54; emphasis added). Woolgar would like to "extend the radical potential of sociological studies of scientific knowledge" (54). He emphasizes that "[i]t is not that science has its 'social aspects', thus implying that a residual (hard core) kernel of science proceeds untainted by extraneous non-scientific (i.e., 'social') factors, but that science is itself constitutively social." It then follows that "we should abandon the idea of science as a privileged or even just separate domain of activity and inquiry" (13).

It is the influence of this kind of perspective on "literature and science" critics which leads chemist and past president of the Society for Literature and Science Stephen Weininger to set as a task for the field, "the quest for meaningful distinctions within the potentially homogenizing discourse of constructivism" (102). Social constructivism has been extremely influential within "literature and science" criticism. Overviewing the field, Weininger notes "the rapidity with which constructivist accounts of science have become not only plausible but respectable" (101). This is hardly surprising. As I have indicated, the recent developments within the philosophy and sociology of science are part of the broad interdisciplinary theoretical movements which also impacted heavily on literary theory; both science studies and post-structuralist literary theory share the realization that all systems of representation are self-referential. Earlier approaches to "literature and science" criticism, by contrast, accepted the assumption that scientific language provides a one-to-one correspondence with reality. As Weininger points out, this effectively exempted scientific discourse from literary analysis: "what could or should a literary analyst do with works that were written in a language that by general assent was understood to be neutral, transparent, free of rhetoric or figuration?" (100). The post-structuralist acceptance of the impossibility of such a transparent language has produced "a torrent of inquiries which are eroding the edifice of classical scientific realism" (101).

To be sure, not all contemporary inquiries within "literature and science" adopt a radically relativistic stance; many critics are intent on formulating an approach that takes into account the relationship between the scientific community and the physical world. Hayles, whose constructivism has been the subject of criticism by members of the scientific community, emphasizes the need to "rescue scientific inquiry from solipsism and radical subjectivism" ("Constrained Constructivism" 84). To reconcile constructivism with the empirical efficacy of science, she proposes a theory of "constrained constructivism," in which scientific theories are understood not as *congruent* with reality (i.e., they do not have a one-to-one correspondence with reality) but rather *consistent* with reality—that is, they *do not contradict* observed reality (80). Similarly, Broderick claims, "What is needed ... to anchor constructivism's scientific perspective ... is something we might call the *insistence of the empirical*" (x). Nevertheless, the influence of the relativist-constructivist view on current "literature and science" criticism can be gauged by Levine's jocular reference to those who do not accept the prevailing consensus as "grouchy realist dissenters" ("One Culture" 13).

One of the strongest connections between science studies and "literature and science" criticism is the assertion that the socially constructed nature of science's knowledge-claims is suppressed or disguised by the conventions of scientific *language*. These conventions operate to erase "any overt traces of both a private and a social process" (Montgomery 13). Scott Montgomery in *The Scientific Voice* lists numerous "grammatical and syntactic strategies" that effect this negation of subjectivity: object-driven narrative; denotative statement achieved by "the banishment of anything overtly resembling literary technique, any concern for sound, rhythm, flourish, and so on"; "minimal use of pronouns and a habitual reliance on transitive verbs"; the absence of "[v]ernacular modifiers, especially those expressive of emotion"; and the use of the passive voice. The result of these rhetorical strategies is that "[w]hat seems to appear is Truth, not a claim for it; the Scientist, not a particular individual; Data, not writing" (13).

Again, this claim—that the scientific style is designed to create a sense of objectivity—hardly comes as a surprise to scientists. It is forcibly argued by Medawar, who claims that "the scientific paper is a fraud in the sense that it does give a totally misleading narrative of the processes of thought that go into the making of scientific discoveries" ("Is the Scientific Paper a Fraud?" 233). The format of the paper, he argues, gives an inaccurate image of its authors' mode of operation: in the "results" section, for example, "You have to pretend that your mind is, so to speak, a virgin receptacle, an empty vessel, for information which floods into it from the external world for no reason which you yourself have revealed" (229). Yet, as before, the claims of some science studies practitioners take observations of this kind to a logical extreme: Woolgar asserts that "representational practices [constitute] the objects of the world" rather than vice versa, and that "science can be construed as a discourse in and through which the prior existence of objects (things) is accomplished" (*Science* 67–8). Woolgar suggests that a scientific report can establish the "'out-there-ness'" of a phenomenon through "isomorphism between textual organization and textual phenomenon." In this sense "there is no object beyond discourse ... the organization of discourse *is* the object" (73).

"Literature and science" critics are often involved in investigating the discursive features that are designed to give authority to science's knowledge-claims. Further, they attempt to show how the qualities constructed by this discourse—objectivity, impersonality, transparency—are effectively undermined by the inevitably figural nature of language. Thus Valerie Greenberg in her analysis of the writings of Max Planck asserts that "self-undermining operations are one of the meanings imparted by Planck's texts to 'scientific'" ("The Scientific Text" 63). Analyses such as Greenberg's, while not necessarily employing the extreme relativist form of constructivism urged by Woolgar, are nevertheless intent on "eroding the edifice of classical scientific realism" (to repeat Weininger's phrase). And this kind of erosion of scientific realism inevitably implies an erosion of the epistemological status of scientific knowledge, and hence of the status of science as a discipline. As Weininger notes, "The constructivist scenario has probably had its greatest impact on our perception of scientific authority. The conviction that science is socially constructed brings with it the realization that scientific authority is constructed as well" (Weininger 101). Given this challenge to authority, it is unsurprising that the relativist-constructivist programme has, for the most part, been perceived by the scientific community as an unwarranted attack.

Popular Science and the "Science Wars"

This challenge to the authority of science from post-structuralist critics brings a new perspective to the "two cultures" debate. Jay Labinger states that "scientists began taking notice [of science studies] in the early 1990s" (204). A 1997 report in *Nature* refers to the "conflict that has been growing in recent years between scientists who argue that science is based on empirical fact, and sociologists who argue that much of scientific knowledge is 'constructed' out of debates between researchers" (Dickson, "The 'Sokal Affair'"). Predictably, the terms of the debate tend to become over-simplified and polarized in this forum. Commonly, social constructivism is presented as purely relativistic in the "judgemental" sense described by Knorr-Cetina and Mulkay: it is, according to another report in *Nature*, "the intellectual school that seeks to describe science as a collection of ideas constructed by one social group—namely natural scientists—and of no greater intrinsic value than any other such collection of ideas" (Macilwain, "'Science and Reason' Forum"). It is this perceived extreme relativism that enrages the scientific community; Cole's brand of "realism-constructivism" is by contrast generally seen as important and useful, particularly as a gadfly in the side of scientism. One *Nature* opinion-piece claims that "an idealistic image of science" is "in many ways ... as dangerous" as a "purely relativistic image" ("Science Wars"). Similarly, physicist David Mermin, writing in *Physics Today*, asserts that "the construction of facts is a subtle mixture of the social and the objective. Sociologists won't get it right by ignoring the latter dimension, any more than scientists can understand the character of their professional activities by ignoring the former" (Letter 15).

Many responses to the work of science critics are more fierce than these reasoned comments. The most prominent example is biologist Paul Gross and mathematician Norman Levitt's 1994 monograph *Higher Superstition: The Academic Left and its Quarrels with Science*. *Higher Superstition* represents a virulent attack on a range of related approaches to "science studies": post-structuralist literary criticism, constructivist sociology, and political criticism including feminist, environmentalist and multicultural analyses. Gross and Levitt pull no punches: in their opinion, constructivist analyses "seem often to escape mere inaccuracy and rush hell-for-leather toward unalloyed twaddle" (43). Gross and Levitt organized a conference under the auspices of the New York Academy of Sciences to address this perceived "'flight from science and reason,'" at which (according to *Nature*) speakers warned that "[social constructivism's] pernicious effects are spreading quickly through universities and (more recently) schools, where they will eventually undermine the standing of science and medicine" (Macilwain, "'Science and Reason' Forum").[1]

Reactions against science studies in the 1990s were not confined to academic forums. As demonstrated in the previous chapter, this was the decade in which the popular science boom reached its peak, and the backlash from the scientific community was often publicly expressed by science popularizers. Labinger identifies the "first significant attacks" on social constructivism by scientists as sections of books by popularizers Steven Weinberg and Lewis Wolpert (204). Weinberg's *Dreams of a Final Theory* features a chapter entitled "Against Philosophy" in which he attacks constructivism; he later published a number of articles on the same theme, in the *New York Review of Books* and other forums (several of these articles are republished in his collection *Facing Up: Science and its Cultural Adversaries*). Wolpert's *The Unnatural*

---

1   The conference, entitled "The Flight from Science and Reason," was held in New York from 31 May to 2 June 1995. The proceedings, edited by Gross, Levitt, and Martin Lewis, were published under the same name.

Nature of Science includes a chapter entitled "Philosophical Doubts, or Relativism Rampant." Several other popularizers have contributed publicly to the debate. As I discussed earlier, Davies sees the attacks by constructivists and relativists as the US equivalent of the British literary snobbishness towards science he so detests. In a 1996 article in the *Sunday Times* he criticizes both anti-science movements, interpreting the "hysterical anti-science tirades" of the British literati and "the mantra of cultural relativism" as evidence of a widening "schism" between the "two cultures" since Snow's time ("The Arts Have Lost It" 13). Richard Dawkins, who in 1995 became the first holder of the Chair of Public Understanding of Science at Oxford University, speaks out against "cultural relativism" in *River Out of Eden* (35–7), *Unweaving the Rainbow* (18–21) and elsewhere ("Postmodernism Disrobed"). Like Davies, he addresses both the "hostility from academics sophisticated in fashionable disciplines" and the anti-science outpourings of Fay Weldon, Bernard Levin and others within an implicit "two cultures" framework (*Unweaving the Rainbow* 18, chapter 2). Even Stephen Jay Gould, while emphasizing that "[s]cience ... is a socially embedded activity" and asserting that "[m]uch of its change through time does not record a closer approach to absolute truth, but the alternation of cultural contexts that influence it so strongly," nevertheless explicitly separates himself from "the overextension now popular in some historical circles: the purely relativistic claim that scientific change only reflects the modification of social contexts, that truth is a meaningless notion outside cultural assumptions, and that science can therefore provide no enduring answers" (*The Mismeasure of Man* 22).

The writers—and thus the readers—of popular science, then, are quite aware of the growing presence of constructivist approaches. Moreover, the debate between the science studies community and the scientific community quickly began to spill over into the non-scientific popular press. The popularization of this academic debate was triggered by physicist Alan Sokal's famous hoax. Sokal submitted to the prominent US cultural studies journal *Social Text* an article entitled "Transgressing the Boundaries: Towards a Transformative Hermeneutics of Quantum Gravity."[2] The article was published in the journal's special double-issue "Science Wars" (Spring/Summer 1996), the aim of which (in the words of its co-editors) was "to gauge how science critics were responding to the attacks by Paul Gross and Norman Levitt and by other conservatives in science" (Robbins and Ross). Sokal's proclaimed goals were to outline "some of the philosophical and ideological issues raised by quantum mechanics and by classical general relativity," to examine "conceptual issues" raised by "the emerging theory of quantum gravity," and to "comment on the cultural and political implications of these scientific developments" ("Transgressing the Boundaries" 201). However, in another article entitled "A Physicist Experiments with Cultural Studies," published in the May/June issue of *Lingua Franca*, Sokal revealed that the *Social Text* article was parodic, and his true intent was to test his

---

2 Sokal's parody and his "Afterword" are repeated in Sokal and Bricmont's extended attack on the misuse of science by intellectuals in the humanities, *Intellectual Impostures* (published in the US as *Fashionable Nonsense*), and all further quotations from these articles refer to this source. The citations of Sokal's *Lingua Franca* article "A Physicist Experiments with Cultural Studies" are taken from the online posting of the article.

suspicions of "an apparent decline in the standards of intellectual rigor in certain precincts of the American academic humanities." Sokal's critique was far more effective than Gross and Levitt's: as a hoax, it had the advantage of dealing a blow that was, in Sokal's words, "self-inflicted" ("A Physicist Experiments"). Moreover, while Gross and Levitt, with their immature insults and reactionary sideswipes at their opponents' "political correctness," laid themselves open to accusations of a right-wing political agenda, Sokal was an active leftist whose proclaimed concern with relativism was its threat to "the values and future of the Left" ("Transgressing the Boundaries: An Afterword" 249).

Although Sokal's attack was weakened by the fact that *Social Text* is not a peer-reviewed journal, the apparent inability of its editors to recognize a parodic imitation of cultural analysis of science nevertheless provoked a highly publicized debate. Sokal's hoax generated international media interest, featuring in front-page articles in the *New York Times*, the *Observer* and *Le Monde*; spurred literary and scientific heavyweights such as Stanley Fish and Steven Weinberg to publicly take sides in the debate, conducted in such literary publications as the *New York Review of Books* and the *Times Literary Supplement*, as well as *Nature* and other scientific journals; and produced "an explosion of postings on academic Internet lists in almost every discipline" (Ross, "Burden"). The esoteric discipline of science studies was suddenly the focus of media attention, with the majority of journalists viewing Sokal's hoax as an indictment of science studies (and cultural studies more generally), and concurring with his opinion that the discipline is characterized by obscurantist language and a proclaimed political radicalism based more on fashionable jargon than effective action.

These often heated debates between science studies practitioners and scientists, which peaked in the 1990s but continue to "simmer" in the new millennium (Baringer 2), have come to be known as the "Science Wars." These "Wars" suggest that a wholesale dismissal of the "two cultures" concept is premature. While constructivism may have shown that science and literature are united in their cultural embeddedness, it has ironically been the source of a renewed, highly public, interdisciplinary squabbling. The debate between constructivism and its opponents has been conducted not solely on the level of epistemology, but has also seen the exchange of sarcastic barbs and implicit slights which have all the signs of two factions at war.

Weapons of Choice in the "Science Wars"

The opponents of constructivism in the "Science Wars" employ a variety of rhetorical tactics. One is to emphasize the degree to which the humanities are governed by fashion rather than logic. Thus Dawkins dismisses cultural relativism in science studies as "a fashionable salon philosophy" and a "voguish fad" (*River Out of Eden* 35; *Unweaving the Rainbow* 18). Another tactic is to stress the imperviousness of science to the babblings of the humanities, which are represented at best as quaintly amusing, at worst as pernicious to the public image of science (and hence science funding), but always peripheral to science itself: "My main concern isn't to defend science from the barbarian hordes of lit crit (we'll survive just fine, thank you)" (Sokal, "Transgressing the Boundaries: An Afterword" 249); "These radical critics are having little or no effect on scientists themselves" (Weinberg, *Dreams* 151); "[Scientists] have heard rumours of literary critics waxing sententious over the uncertainty principle, or Gödel's theorem; but if so, they have almost certainly written these efforts off as harmless, even charming, examples of the literary temperament" (Gross and Levitt 4).[3] A more insidious attack is the suggestion that constructivism is just a case of physics-envy gone sour; that disciplines such as sociology, having failed in their struggle to become scientific, are attempting to give themselves an ego-boost by denying the epistemological uniqueness of scientific knowledge: "too great an aspiration can lead to frustration. ... what hope is there for sociology acquiring a physics-like lustre?" (Wolpert, *Unnatural Nature* 121). Another angle is to emphasize the lack of concrete scientific knowledge on the part of science studies practitioners, as Gross and Levitt do; furthermore, they relegate those supporters of cultural constructivism who do have scientific credentials to "the movement's fringes," along with "the occasional refugee from an unsatisfactory scientific career" (6). The implication is that scientifically trained critics are, like sociologists, spurred on by a desire to destroy what they could not themselves achieve: scientific authority.

The critics at whom these shots were fired retaliated with their own fair share of intellectual condescension. The editors of *Social Text*, Bruce Robbins and Andrew Ross, in their online response to Sokal's claim that his article was a parody, are quick to deny that they originally saw intellectual merit in the article. They considered it "a little hokey"; Sokal's "adventures in PostmodernLand" were not their "cup of tea." They go on to claim that Sokal's article "would have been regarded as somewhat outdated if it had come from a humanist or a social scientist"; as the work of a scientist, it was "unusual," and "plausibly symptomatic of how someone like Sokal might approach the field of postmodern epistemology, i.e., awkwardly but assertively trying to capture the 'feel' of the professional language of this field, while relying upon an armada of footnotes to ease his sense of vulnerability." Both Robbins and Ross's language and their implied argument—that the *Social Text* editors will accept articles they consider substandard if these articles are written by *scientists*—are patronizing. Robbins and Ross also challenge Sokal on political grounds, questioning "his own good faith as a self-declared Leftist." Elsewhere Ross claims generally that hidden political agendas lie behind attacks on science studies: "The erosion of the cold war funding contract with the state, combined with the decrease in public respect for scientific authority, has created a demand for scapegoats in the demonic form of politically motivated scholars in science studies" ("Burden").

Thus scientists are portrayed as greedy and power-hungry, just as critics in the humanities are viewed as suffering from intellectual inferiority complexes leading to

---

3   The impact of these slights on the literary community is amplified when one realizes that this kind of condescension is not limited to critics of science studies; even Gould, constructivist literary critics' popularizer of choice, once referred to the Society for Literature and Science as a "sweet little society" (qtd. in Ackrill 136).

physics-envy. It is important not to dismiss these accusations as nothing more than petty name-calling;[4] yet, as new expressions of old disciplinary grievances, they did little to assuage Robbins and Ross's concern that science studies would be perceived simply as the site of "an academic turf war between scientists and humanists/social scientists." It is difficult to determine whether this hostile situation represented a process of diagnosis of the "two cultures" concept, or a reinscription of it. The original "two cultures" divide in a sense moved to one side of the new debate: in the "Science Wars," one side (made up predominantly of those working in the humanities and social sciences) argued for a form of disciplinary unity, with science and literature both forms of "discourse" with equal epistemological validity—"one culture"; and the other (predominantly natural scientists) argued for the separateness of science, the uniqueness of its knowledge-claims—"two cultures" (at least in an epistemological sense).

One feature of Snow's "two cultures" model that recent debate does appear to have reinforced is the lack of interdisciplinary communication, manifested in a tendency towards polarization and caricature in each discipline's representation of the other. As I have mentioned, the primary objection voiced by the popularizers and other scientists quoted above is to the more extreme relativist versions of constructivism. Weinberg, for example, welcomes the "useful historical and sociological observations" provided by critics, but insists, "It is simply a logical fallacy to go from the observation that science is a social process to the conclusion that the final product, our scientific theories, is what it is because of the social and historical forces acting in this process" (*Dreams* 148; 149). A common rebuttal offered against the constructivist programme is the material efficacy of science. Colin Bruce's popularization, *The Strange Case of Mrs Hudson's Cat*, which appropriates Arthur Conan Doyle's detective Sherlock Holmes as a means to explain relativity and quantum physics, features a character who supports the belief that "the world view of, say, an American Indian or an Australian Aborigine [has] as much right to be respected as yours or mine," and that one should not "deny the validity of their cultures." Holmes wonders if his client would espouse a similar view if at sea, short of rations, with a flat-earth navigator: "I think that might concentrate his mind wonderfully as to the validity of different world views! Ah, the conceit of the artistically minded aristocrat" (7). Holmes's jibe may be directed at his client, but Bruce's target is presumably another "artistically minded" group of people. Dawkins phrases the familiar objection more directly: "Show me a cultural relativist at thirty thousand feet and I'll show you a hypocrite. ... If you are flying to an international congress of anthropologists or literary critics, the reason you will probably get there—the reason you don't plummet into a ploughed field—is that a

---

4   One must acknowledge, for example, the relevance of Ross's accusation, given the concern expressed by some critics of science studies that a central "danger" it presents is its impact on "those in charge of funding science" (Weinberg, *Dreams* 151). Conversely, Levine effectively supports the idea of "physics envy" when he admits that one of the motivations behind philosophers' and literary critics' eagerness to dispel scientific realism is "a pleasure in the deflation of scientific authority" ("One Culture" 15).

lot of Western scientifically trained engineers have got their sums right" (*River Out of Eden* 36).

This argument seems convincing; but, as should be clear from the previous discussion, it relies on a reading of constructivist theory that many critics would deny. Dawkins, while admitting that there are a range of views within "cultural relativism," states that the "extremist" view he describes is "alarmingly common" (*River Out of Eden* 36). Yet, even remaining within Dawkins's own example of gravitational attraction, it is easy to cite critics in the humanities who essentially agree with him: "However socially constructed Newton's theory of gravitation was," argues Broderick, "it would scarcely have had great success if (influenced, let us say, by the Holy Trinity) it had proposed an inverse cube law" (87). Hayles agrees that while "one can imagine" explanations of gravity other than the Newtonian or Einsteinian theories—for example, "a Native American belief that objects fall to earth because the spirit of Mother Earth calls out to kindred spirits in other bodies"—nevertheless "no viable model could predict that when someone steps off a cliff on earth, she will remain spontaneously suspended in mid-air. This possibility is ruled out by the nature of physical reality" ("Constrained Constructivism" 79). An even more startling counter-example can be found in Knorr-Cetina's theorizing; more than ten years before Dawkins's comment, she pre-empted his reasoning exactly, describing a typical argument against relativism as one which

> challenges relativism on the ground that its proponents surely would not attempt to walk through a wall, or … attempt to step out of a window on the thirtieth floor in order to get to the street level. The implication here is that relativism denies the existence of a material world, and that this denial is prudently but somewhat hypocritically suspended in relativists' everyday behaviour. ("The Constructivist Programme" 320)

This challenge, Knorr-Cetina argues, disappears when one accepts the difference between epistemic and judgemental relativism. However convincing this distinction, it is clear from his ignorance of or indifference to these arguments that Dawkins, if he does not consciously misrepresent his opponents, is not willing to pursue the debate on any but the most superficial terms.

Dawkins's strategy is one which Robbins and Ross identify in the attacks of Sokal and Gross and Levitt: that is, the presentation of constructivist critics as "caricatures in the form of otherworldly fanatics who deny the existence of the facts, objective realities, and gravitational forces" (Robbins and Ross). Similarly, Anthony Gottlieb, in a reply to Wolpert's dismissal of philosophy of science, emphasizes that "[t]he overwhelming majority of philosophers of science are not relativists" (Wolpert and Gottlieb 18), and argues that an uncritical acceptance of the "journalistic hyperbole" which highlights the "outrageous fringes" of philosophy of science is "rather like taking Rupert Sheldrake as a paradigm example of a scientist" (17). Philip Kitcher agrees, claiming in his "Plea for Science Studies" that the extreme examples cited by critics of science studies often originate from practitioners who are not central to the field (53 n. 20).

The same strategy can be identified on the other side of the disciplinary debate, with post-structuralists attacking a version of science that in reality has very few serious supporters. Terry Eagleton claims, "The model of science frequently derided by post-structuralism is usually a positivist one—some version of the nineteenth-century rationalistic claim to a transcendental, value-free knowledge of 'the facts'. This model is actually a straw target. It does not exhaust the term 'science', and nothing is to be gained by this caricature of scientific self-reflection" (144). Conversely, it also appears that the proponents of the relativist-constructivist viewpoint overstate their *own* case for rhetorical purposes. Knorr-Cetina describes epistemic relativism as "a very cautious epistemological perspective which is *primarily directed against the bolder doctrines of epistemic realism*" ("The Constructivist Programme" 321; emphasis added); elsewhere she and Mulkay write that "*insofar as scientific realism wishes to make such a claim* [that scientific knowledge mimics nature], epistemic relativism is anti-realist" (5; emphasis added). Cole notes, "An intellectual movement which takes a position at variance with existing beliefs is frequently forced into overstating its claims in order to differentiate itself from the position it argues against" (x). The phrasing of these definitions of epistemic relativism lends support to his claim that the proponents of relativism-constructivism are primarily concerned with combating an existing view. Because of previous assumptions that science was off-limits to sociologists as a subject of investigation, constructivists "remained fearful that if nature is even included as one of several influences on the cognitive content of science, then social influences will be seen as secondary or unimportant" (Cole x). Cole's view of the motivation behind extreme versions of constructivism is remarkably similar to Eagleton's description of the situation that developed when post-structuralist criticism became "a fashionable style in Left academic circles":

> To employ words like "truth'" "certainty" and the "real" was in some quarters to be instantly denounced as a metaphysician. If you demurred at the dogma that we could never know anything at all, then this was because you clung nostalgically to notions of absolute truth, and to a megalomaniac conviction that you, along with some of the smarter natural scientists, could see reality "just as it was." (144)

Eagleton argues the same general point as Searle and Cole when he insists, "To say that there are no absolute grounds for the use of such words as truth, certainty, reality and so on is not to say that these words lack meaning or are ineffectual" (144).

Thus, it appears that both constructivist critics and those who defend science against constructivism often resort to building their arguments around a caricature of their opponents' view. This is unsurprising; for while the theoretical issues concerned are highly abstract, their repercussions are quite concrete. What is primarily at stake in the current debate is the nature of scientific authority, a concept that is inextricably tied to questions of disciplinary hierarchy, research funding, social and career status, and political ideology. It is not my aim to choose between a traditional and a relativistic theoretical perspective for this discussion of physics popularization—as Cloître and Shinn suggest, it is doubtful if either of these models "adequately describes or accounts for the complexities of contemporary expository practices" (31). My purpose in including an outline of these current theoretical concerns is to enable the question on which the remaining chapters of this book are focused: what perspective on cross-disciplinary debates can an analysis of popular physics books provide?

Popular Science as Battle Ground

Before exploring this question, one might ask *why* physics popularizations (as a subset of science popularizations more generally) should be expected to impinge on these debates. One reason is timing: the chronology of the "Science Wars" runs parallel with that of the popular science boom, as described in the previous chapter. As I have outlined, the early beginnings of the "Wars" can be traced to the 1970s with the growth of SSK, but they reached their zenith in the 1990s. And like the boom, the "Wars" had begun to die down by the turn of the new millennium. The opening of the twenty-first century saw the publication of two multi-disciplinary essay collections with explicitly conciliatory aims, as indicated by their titles, *After the Science Wars* and *The One Culture? A Conversation about Science*. Talk began to turn cautiously towards "peace"; essay titles in the latter collection include phrases such as "Beyond War and Peace," "Science Peace Process" and "Peace at Last?" Both collections look towards more productive, considered debates (Baringer 2; Labinger and Collins, Preface ix), just as those who observe the demise of the popular science boom hope that the former "flood" of books will become a "small but higher quality trickle" (Evans). While the possible causal connections between the boom and the "Science Wars" are complex and debatable, this shared timing certainly meant that popularizers became important participants in the "Wars" (and later in the nascent "peace efforts"). In some cases this involvement was quite deliberate and acknowledged: popularizers such as physicist Steven Weinberg explicitly criticized aspects of science studies in their books and in the press. Perhaps more important, however, are the cases in which issues related to the "Science Wars" are dealt with, implicitly and possibly subconsciously, in the structure of popularizations themselves.

Several commentators have noted the tendency for popularizers to produce naïve realist accounts of scientific knowledge which automatically preclude any constructivist perspective. David Hess begins his discussion of social constructivism by opposing it to "[p]opular accounts" which "sometimes assume that science and technology are produced according to purely rational (or cognitive) factors" (2). George Johnson, himself a science writer, is more trenchant: "Almost all science books, popular and unpopular, are written on the assumption that there actually are laws of the universe out there, like veins of gold, and that scientists are miners extracting the ore. We are presented with an image of adventurous explorers uncovering Truth with a capital T" (5). Hayles similarly observes that, despite the advent of constructivism, the notion that scientific knowledge accesses reality in a direct way "continues to have a vigorous existence in popular culture as well as in the presuppositions of many practicing scientists." In popular culture, she argues, this view is heavily reinforced by the style of popularizations:

> When the TV camera, accompanied by Carl Sagen's [*sic*] voice-over, zooms through the galaxy to explore the latest advances in cosmology, these presuppositions are visually and

verbally encoded into an implied viewpoint that seems to be unfettered by limitations of context and free from any particular mode of sensory processing. As a representation, this simulacrum figures representation itself as an inert mirroring of a timeless, objective reality. ("Constrained Constructivism" 79)

Montgomery in *The Scientific Voice* notes that one standard set of tactics adopted by science writers centres on seeming merely to "interpret" science neutrally: like the scientific report writer, this kind of popularizer pretends "no involvement in the historical and institutional conditions for knowledge, nor in its social implications" (49). These writers fail to admit that "[n]ot merely a source or mediator, science writing is also a creator of social reality, both for those within and those without science" (48).

While these observations refer to popularizations of all scientific disciplines (the present study could have looked at popularizations of biology or another science with fruitful results), physics and popular physics books have had a special role in the "Science Wars." As I discussed in my introduction, physics has been traditionally represented as the "purest," "hardest," least subjective and hence least socially embedded science, the "gold standard" to which other sciences are supposed to aspire. Weinberg, in his article "Peace at Last?" in the collection *One Culture*, states that at least one issue remains to be debated in the "Science Wars": "Although scientists recognize that their theories often bear the stamp of the social environment in which they are formulated, we like to think of this as an impurity, some slag left amid the metal, which we hope eventually to eliminate." In a footnote, however, he qualifies the term "scientists": "Well, at least physicists. Or at least some physicists. Or maybe just me" (239). There is still a sense, for Weinberg at least, that physics remains the ultimate battleground in the "Science Wars." It is perhaps no coincidence that, as Labinger and Collins report in their introduction to *The One Culture?*, "the majority of scientists who have been paying serious attention to science studies have been physicists" (8), and that two of the most vocal scientists in the "Wars"—Weinberg and Sokal—are physicists. It is hard to imagine that Sokal's hoax article would have been so readily accepted in the press as patently ridiculous if it had focused on the "cultural and political implications" of evolutionary biology rather than quantum theory. The reputation of physics (and physicists) as science's "gold standard" means that physics popularizations are particularly interesting to examine within the context of recent cross-disciplinary exchanges.

Most of the popularizations I have mentioned in this chapter directly address the debates of the "Science Wars." The texts examined in the following chapters make no explicit interventions in this battle; their strategies are less obvious, but no less influential. Like all popularizations, they produce their own versions of scientific knowledge, and hence of scientific authority. Thus, automatically, they assume some perspective on cross-disciplinary issues of the kind discussed in this chapter: how scientific knowledge can be related to language, reality and other forms of knowledge. Disciplinary relations are thus entangled in the structure and style of these popular science books. The next three chapters are devoted to analysing the "novelistic" features of physics popularizations—figurative language, narrative structure, characterization—and exploring the implicit view of science they encode.

CHAPTER 4

Knowing Quanta: Anthropomorphic Metaphor in Popularizations of Quantum Theory

Few branches of science have proven more readily and usefully adaptable to the interests of literary critics than quantum theory. This previously obscure field experienced a burst of popular interest in the second half of the 1970s, following the successes of New Age popularizations *The Tao of Physics*, by Fritjof Capra, and *The Dancing Wu Li Masters*, by Gary Zukav. Since that time, numerous quantum phenomena have been assigned literary parallels: "complementarity corresponds to the oxymoron, holism to the synecdoche, fusion of identities to the metaphor and the shifting of identity to the metonymy. The observer-participation has its counterpart in the problem of interpretation" (Vanderbeke 253–4). Just as, according to Heisenberg's Uncertainty Principle, "the very act of observation influences the observed results," so in certain literary texts "the act of communication influences the concepts and descriptions being communicated" (Booker 581). The wave-particle duality is in one critic's view a challenge to "the Aristotelian logic of identity that informs all Western metaphysics" (Best 199); for another, it brings a new perspective to theories of mimesis and semiosis: "Mimesis, because it emits bits of new information, resembles a corpuscular description, and semiosis, because it repeats the same information or an invariant theme, is wavelike" (Sporn 212). Even the esoteric Einstein-Podolsky-Rosen (EPR) thought experiment can be appropriated for literary-theoretical purposes, as "[p]erhaps the most unsettling evidence of the invisible (or 'occulted') presence of the other, or the apparently absent" (Froula 303).

As these examples indicate, one of the standard methodologies of "literature and science" critics in the last few decades has been the construction of analogies between literary and scientific concepts. This analogical or metaphorical method has been a topic of contention within the "Science Wars." In his hoax article, Alan Sokal lists among his copious notes numerous examples of "the cross-fertilization of ideas between relativistic quantum theory and literary criticism" ("Transgressing the Boundaries" 202 n. 9), and it is analyses such as these that are the targets of his parody. His article constructs implied parallels between quantum theory and critical theory that he retrospectively identified as spurious, deliberately deflating the presumed enthusiasm of his readers (and *Social Text*'s editors), and bringing the validity of this method into question.¹ Unsurprisingly, Sokal and co-author Jean Bricmont raise the issue of metaphor and analogy early in their follow-up to the hoax, *Intellectual Impostures*. They are careful to point out that they have no grievance with poets or novelists who use scientific terms and ideas "out of context and without really understanding their scientific meaning" (8), but are less forgiving when it comes to critics and theorists. Noting that "a metaphor is usually employed to clarify an unfamiliar concept by relating it to a more familiar one," they question why theorists would want to explain concepts in their own field in terms of another esoteric and technical field, suggesting that the real goal might be to give a sense of profundity to otherwise banal observations (9). But metaphors and similes do not always express the unfamiliar in terms of the familiar (is the evening sky less familiar than an etherized patient?), nor are they always pedagogical in purpose. The question of what "slippage" occurs when analogies are constructed between scientific and literary concepts, and what the consequences of that slippage might be, requires a far more complex approach.

One of the complexities of examining the analogies that critics construct between literature and science lies in the metaphorical richness of popularizations themselves. Many literary critics initially engage with scientific ideas through these texts, and thus through the metaphors popularizers employ. Yet, ironically, when "literature and science" critics turn to popularizations for information, they often appear to forget the status of these books as textual constructions, treating them as straightforward, transparent summaries of what is assumed to be a homogeneous and harmonious scientific community. This means that metaphors (and other literary devices) employed by popularizers to achieve particular ends are accepted at face value rather than interrogated, with a consequent impoverishment of cross-disciplinary exchange.

The uncritical use of science popularizations as information sources by researchers in non-scientific fields is something to which philosophers of science have recently drawn attention. Stephen Kellert, in his article "Extrascientific Uses of Physics," examines legal theorists' metaphorical uses of chaos theory. While Kellert defends such uses, he is concerned about problems they engender, including the inaccuracies produced by "an over-reliance on popular accounts of the science" (S458). Cathryn Carson looks more broadly at claims by critics in the humanities (and particularly literary critics) that chaos theory, quantum theory and relativity make up a "postmodern physics." She challenges such claims on two main grounds. First, she notes that arguments for a "postmodern physics" have been "about parallels—about analogues, about consonances, about correlates"; she raises questions about "the meaning of parallels" (642). Secondly, she points out

---

1   In this sense, Sokal's hoax is reminiscent of a telling cross-disciplinary incident described by Richard Feynman in one of his many anecdotes. Feynman relates how, as a young man, he attended a talk in which a speaker suggested an analogy between mathematics and poetry. Feynman delighted the speaker by extending the analogy to theoretical physics, and then deflated him by declaring such analogizing loose and unmeaningful (*"Surely You're Joking"* 66). See Gossin (92) for an analysis of this incident within a "literature and science" context.

that critics' reliance on popularizations of physics mean that they remain within the framework constructed by these popularizations, and merely replicate the debates established there (650). In particular, she locates the "philosophical" subgenre of physics popularizations on which critics rely as one which began with the New Age popularizations of the 1970s (648–9).

These analyses suggest that it is the textual strategies of popularizers on whom critics in the humanities depend, as much as the strategies of those critics themselves, that require examination. This chapter focuses on the strategic slippage in meaning that can occur when popularizers use metaphors to advance an argument. It concentrates on a specific subset of physics popularizations, those dealing with quantum theory. While all scientific fields rely on metaphor in both professional and expository contexts, the relationship between quantum theory and metaphor is a particularly problematic one, and thus a particularly productive one to study. I begin by outlining theories of metaphor and scientific language, before focusing more specifically on the use of metaphor in quantum physics popularizations. In particular, I examine the problems surrounding the use of anthropomorphic metaphor, in the context of frequent claims that quantum mechanics has heralded a return to an anthropocentric world-view. I conclude with an analysis of the use of anthropomorphic metaphor in a popularization particularly favoured by literary critics, Zukav's *The Dancing Wu Li Masters*. Zukav's book has been highly effective in stimulating the interest of non-scientists in quantum theory. However, I contend that its usefulness is countered by its dangerously loose metaphors, which exacerbate the kinds of communication problems that have fuelled the "Science Wars." My wider argument is that constant attention to the textual strategies of science popularizations is a prerequisite for productive cross-disciplinary research and debate.

Metaphor in Scientific Discourse

The relationship between metaphorical language and scientific discourse is central both to recent theories of metaphors and to the arguments about the nature of scientific and literary knowledge discussed in Chapter 3. Andrew Ortony opens his general introduction to the anthology *Metaphor and Thought* by noting, "Science is supposed to be characterized by precision and the absence of ambiguity, and the language of science is assumed to be correspondingly precise and unambiguous—in short, literal" ("Metaphor" 1). He opposes this approach to one in which "any true veridical epistemological access to reality is denied," a view which "provides no basis for a rigid differentiation between scientific language and other kinds." Ortony broadly terms these two approaches to language "nonconstructivism" and "constructivism," and relates them to two approaches to metaphor. The constructivist view "tends to undermine the distinction between the metaphorical and the literal," as it considers all language non-literal. In the non-constructivist view, by contrast, metaphors are "deviant, and parasitic on 'normal usage'"; metaphors "characterize rhetoric, not scientific discourse." These are, as Ortony admits, extreme positions, which are adopted "to different degrees, and in different ways" by various scholars (2).

As in previous chapters, it is not my intention here to untangle the constructivist/non-constructivist distinction. Damien Broderick's claim that "even the most rigorous piece of mathematised physics" relies either on metaphor—"the evolving equation somehow *resembles* the entire phenomenon"—or metonymy—"the equation mimics a crucial *part* of the phenomenon, which stands for the whole" (90)—is persuasive, but also, in my opinion, trivial (in the mathematical sense). William Paulson has argued that the ease with which the familiar deconstructionist claim that "all language is always already literary" is made is "an index of how little [the claim] would actually tell us" (63–4). Paulson asserts, "What we must deal with are an ensemble of properties that enable us to speak of texts as more or less rhetorical, more or less poetic, without claiming that such a thing as a nonrhetorical or completely unpoetic text exists in natural language" (64). Similarly, epistemological egalitarianism renders the term "metaphor" redundant, unless one accepts that while all statements might be considered metaphorical in some sense, some statements are nonetheless more metaphorical—or, to use Max Black's term (25), more "actively" metaphorical—than others.

The nature of metaphor in general, and its role in science in particular, has been the focus of a large body of multi-disciplinary research, employing a diverse range of competing terms and definitions; any analysis must build its framework from a limited selection of these. For literary critics, metaphor is defined most simply as a figure of speech that constructs an "imaginary identity" between two subjects, to suggest "some common quality shared by the two" (Baldick 134). The two subjects are often distinguished using I. A. Richard's terminology: the "principal" or literal subject is the "tenor," the figurative one the "vehicle" (96–7). Related terms include the simile, in which similarity, rather than identity, is explicitly identified as the basis for the relationship between the two subjects; and the analogy, which can be considered a kind of "predictive metaphor" that functions by "ranging two patterns of experience alongside each other, seeking their points of identity, then using one pattern to extend the other" (Beer, *Darwin's Plots* 74). As this last definition indicates, literary critics (myself included) tend to use "metaphor" as an umbrella term which includes similes and analogies as well as metaphors in the narrower sense defined above.

While traditionally critics understood metaphor as a rhetorical ornament, a figurative expression of a comparison that could be paraphrased literally, this view was challenged in the twentieth century. In *The Philosophy of Rhetoric* (1936), Richards argues that metaphors depend as much on the distance and differences between the two subjects (they are held in "tension") as the similarities (or "ground") between them (117; 125; 127). He also claims that in metaphor the vehicle and tenor interact in such a way that a new meaning is created, which cannot be reduced to a "plain meaning"; and that this creation of meaning can change the reader's understanding of the vehicle as well as the tenor (100). Richard's ideas have been developed into the influential "interaction" view of metaphor by Max Black. Black emphasizes the fact that metaphors require readers/hearers to *select* the properties of the vehicle (Black uses the term "secondary subject") that fit the tenor (or "primary subject")

(28). Thus, to use Black's example (28–9), in the metaphor "Marriage is a zero-sum game" the secondary subject includes ideas of contest, opponents, and winning at another's expense, but it is not entirely clear exactly to what degree, and in what sense, each of these properties applies to the primary subject, "marriage"; the reader must make these connections. Hence, "we cannot set firm bounds to the admissible interpretations: Ambiguity is a necessary by-product of the metaphor's suggestiveness" (30). In this sense, metaphor is not merely a summary of "a literal point-by-point comparison," as the latter lacks metaphor's ambiguity and suggestiveness. Metaphors cannot be pinned down, Black argues, as can more direct comparison (in which he includes analogies and similes). Further, Black holds that there is no "infallible test" for "discriminating the metaphorical from the literal"; a particular statement could be read as metaphorical or literal, depending on reasons such as its "congruence" (or lack thereof) with the text surrounding it (34). Again, it is the reader who must make this judgement.

Philosopher of science Richard Boyd uses Black's observations to launch his analysis of the role of metaphor in science ("Metaphor and Theory Change"). He notes in particular Black's view that metaphor, with its ambiguity and suggestiveness, is most useful where scientific precision is not possible. In this view, which corresponds to the "nonconstructivist" position described above, metaphor cannot constitute a core part of scientific theorizing. There remains, however, a role for heuristic metaphors in science—those that facilitate the process of investigation; and pedagogical or exegetical metaphors—those that "play a role in the teaching or explication of theories" but are not integral to the theories themselves, which have other, "nonmetaphorical (or, at any rate, less metaphorical) formulations" (482; 485). As an example of the latter, Boyd cites "the description of atoms as 'miniature solar systems'" (486). However, he argues that science additionally utilizes "theory-constitutive" (or what elsewhere are termed "generative") metaphors that, unlike heuristic or pedagogical metaphors, are "an irreplaceable part of the linguistic machinery of a scientific theory." He gives as an example the claim that "the brain is a sort of 'computer'" in cognitive psychology (486). More recently Susanne Knudsen, in an article entitled "Scientific Metaphors Going Public," has used Boyd's categories to explore the role of metaphors in science popularization, which she identifies as a neglected topic within the field. Through an empirical study focusing on the metaphors of "code" and "translation" in molecular biology, Knudsen shows that the distinction between pedagogical and generative metaphors is not clear-cut; the same metaphor can have a different function in two different texts. Like Black, Knudsen emphasizes the importance of context: Boyd's categories depend on not "the specific metaphorical expression itself, but on the context and its purpose" (1261).

Literary critics interested in science writing, such as Gillian Beer, have also emphasized the way the context of a metaphor determines the extent to which its ambiguity can be exploited. Some vehicles, for example, resonate with broader themes, both outside and within the text, in a way that opens up otherwise dormant interpretations. This is most obvious perhaps in popular discourses of scientific disciplines that have direct identifiable social repercussions, such as sociobiology; but is also quite evident in the language of popular physics. One obvious example is the "space is an empty ocean" metaphor I briefly mentioned in Chapter 1. The primary purpose of this metaphor is pedagogical: the reader is encouraged to identify certain physical characteristics of the empty ocean with those of space. However, this metaphor takes on extra resonance and exceeds its pedagogical function when considered in the context of space exploration and potential colonization: the reader might identify social and historical characteristics of the ocean, such as its role in the expansion of empire, with space, and thus see space as primarily a means to reach and conquer new frontiers. Another example is Alan Guth's dubbing of his theory of the exponential expansion of the early universe "inflation," a term which resonated with wider cultural trends during the 1980s. Furthermore, when vehicles are calculated to exploit a particular social trend, the possibility opens for this trend to be naturalized through its metaphorical projection onto a physical phenomenon. While it is stretching a point to suggest that Guth "naturalized" the concept of inflation, Greg Myers has argued a similar case more convincingly, suggesting that Stewart and Tait's use of a capitalist metaphor to explain physical principles in their popularization *The Unseen Universe* also works in reverse, exploiting the reader's knowledge of these principles to reinforce capitalist economic theory ("Nineteenth-Century Popularizations" 55–7). As the "interaction" model of metaphor suggests, the change in perspective enacted in the metaphor operates on both the vehicle and the tenor.

Often vehicles that are apparently chosen for neutral qualities such as size or structure will carry mythic or religious connotations. In Tom Stoppard's play *Hapgood* a physicist likens the atom to St Paul's and an orbiting electron to a "moth in the empty cathedral" (49). This metaphor has been singled out by one literary critic for praise: "Other structures might be to scale, but concert halls, train stations, or sports arenas would not adequately convey [the physicist's]—and Stoppard's—sense of awe at the mysteries of the quantum world" (Delaney 127–8). However, these connotations did not originate with Stoppard, as the metaphor is, in various forms, a staple of physics popularizations. Capra has the cathedral as St Peter's in Rome and the moths as specks of dust (5); Zukav states that "the electrons orbiting the nucleus are 'like a few flies in a cathedral,'" attributing the metaphor to Rutherford (38). Conversely, the fact that John Gribbin uses this metaphor at least twice but transforms the cathedral into a secular structure on both occasions—Carnegie Hall in one case (*In Search of the Big Bang* 146) and Albert Hall in the other (*In Search of Schrödinger's Cat* 260)—can be read as an attempt to quash these religious connotations. In summary: there are certain metaphors and analogies which, due to their embeddedness within a wider context, encourage the reader to apply properties of the vehicle to the tenor (and vice versa) other than those strictly required for pedagogical purposes.

Quantum Theory and Metaphor

The possibilities and problems created by extra-pedagogical connotations of metaphors are relevant to any kind of science popularization. However, these issues are particularly pertinent to popularizations of quantum theory, which occupies a unique position with respect to figurative language due to the non-sensory and seemingly nonsensical nature of subatomic phenomena. For example, consider the examples mentioned above in which an electron in an atom is likened to a moth or a fly in a cathedral or concert-hall, or (in Boyd's example) to a planet moving around the sun; while these analogies convey useful structural information, they all reinforce a *particulate* conception of subatomic entities. Yet fundamental to quantum theory is the realization that electrons are not spatially contained in this way. As Thomas Kuhn points out, even Bohr's ostensibly non-metaphorical model of the atom, in which the nucleus and the surrounding electrons are "represented by tiny bits of charged matter," nonetheless assumed a "metaphorlike process. Bohr's atom model was intended to be taken only more-or-less literally; electrons were not thought to be exactly like small billiard or Ping-Pong balls …" ("Metaphor in Science" 538). The issues this raises for popularizers were identified early in the development of quantum theory. J. W. N. Sullivan in 1925 pointed out the difficulties experienced by expositors of this period, who were used to employing "pictorial" models, the standard tool of nineteenth-century popularizers: "The method is still successful in many expositions. Thus an atom is presented as a 'miniature solar system' with the result that ordinary readers have no difficult in understanding atomic theory—provided the quantum theory is not mentioned. A way of presenting quantum theory pictorially has not yet been invented" (*Three Men* v). In the standard Copenhagen interpretation of quantum theory developed by Bohr and Heisenberg, the question of what electrons actually are is explicitly avoided.[2] Far from conforming with the classical view in which each component of reality has a one-to-one relationship with a component of scientific description, quantum theory cannot claim to refer to reality, but only to the results of experiments (Squires, *Mystery* 117–18). Similarly, the mathematics employed in quantum theory does not provide a model for a system, but is better thought of as a tool to predict experimental outcomes.

In one sense, then, the quantum world, foreign as it is to our everyday terms and categories, can only be spoken about in metaphor. In another sense, it implies a radical failure of metaphor: the metaphors popularizers employ are all vehicle and no tenor. Eddington suggested as much in his popularization *The Nature of the Physical World*: "No familiar conceptions can be woven round the electron … . *Something unknown is doing we don't know what*—that is what our theory amounts to." For Eddington, nonsense verse provides the most appropriate metaphorical description for the quantum domain: "Eight slithy toves gyre and gimble in the oxygen wabe; seven in nitrogen. By admitting a few numbers even 'Jabberwocky' may become scientific." He concludes, "It would not be a bad reminder of the essential unknownness of the fundamental entities of physics to translate it into 'Jabberwocky'; provided all numbers—all metrical attributes—are unchanged, it does not suffer in the least" (291).

Thus quantum theory, as well as challenging classical notions of causality and precision, also entails a rethinking of the nature of descriptive language. This was clear to the founders of the theory. "Can one speak about the atom itself?"

---

2 Except where otherwise stated, throughout this chapter when I refer to "quantum theory" I am assuming the Copenhagen interpretation.

Heisenberg asks in *Physics and Philosophy*, adding, "This is a problem of language as much as of physics ..." (146). Of course, scientists had been dealing with microscopic phenomena that were not directly apprehensible for many years before the advent of quantum theory. Beer notes Maxwell's difficulties in describing molecules in a paper which he gave to the British Association for the Advancement of Science in 1873: "His problem is how to describe the new concept, 'molecule', since 'no one has ever seen or handled a single molecule' and molecular science is 'one of those branches of study which deals with things imperceptible by our senses, and which cannot be subjected to direct experiement [*sic*]'" (Beer, *Open Fields* 150). However, before quantum theory, physicists could at least consider a molecule or an atom as a tiny ball, or some combination, static or dynamic, of tiny balls. Although this kind of metaphorical picture is immediately problematic (what is the ball made up of?), it does not require the total abandonment of macroscopic concepts in the way that quantum theory, with its wave-particle duality and indeterminacy, insists upon.

A significant consequence of the inability of quantum physicists to formulate a one-to-one relationship between "reality" and language has been a conscious shift away from attempts at univocality towards a more "literary" use of language. Beer notes the "contrary mode of impressionistic or whimsical naming which is fashionable in high theory today: words such as 'charm', 'quark', or 'black hole' deliberately evade severe equivalence in order to allow space for correction and enhancement without the need constantly to replace and to move on from terms" (*Open Fields* 157). Terms like "charm," then, are essentially vehicles designed to avoid strict attachment to a tenor. Heisenberg noted this trend toward figurative language in his *Physics and Philosophy*: describing the "language concerning the atoms [which] has actually developed among the physicists in the thirty years that have elapsed since the formulation of quantum mechanics," he observes that the complementarity principle encouraged "an ambiguous rather than an unambiguous language" (154), and declares this trend "in many ways quite satisfactory, since it reminds us of a similar use of language in daily life or in poetry" (155). Thus, in contexts where mathematical language is inappropriate, metaphor—a device in which two concepts are held together in tension—might be the best way of conveying the paradoxes and dualities of quantum theory. However, as discussed above, metaphor is also a highly slippery tool: the flip side of its convenient ambiguity is its tendency to take on unintended connotations, or, alternatively, to be read overly literally.

Popularizers, unable to describe the quantum world in any concrete way, make extensive use of metaphor. One literary technique that popularizers find particularly useful is anthropomorphic metaphor (or "personification")—the metaphorical attribution of human qualities to non-human phenomena. Anthropomorphic metaphor is a device that raises its own set of issues in science popularization, depending on the particular scientific field in question. In popularizations of animal behaviour, for example, the problematic potential of anthropomorphism is fairly apparent.[3] The dangers of anthropomorphism in accounts of quantum theory might appear fewer, as the distance between the quantum world and the

---

3 See D. R. Crocker's article "Anthropomorphism: Bad Practice, Honest Prejudice?" for a brief discussion of anthropomorphism in both popular and professional accounts of

human world would seem likely to prevent the misinterpretation or literalization of metaphor. This situation changes, however, in popularizations that attempt to draw links between quantum theory and human consciousness, or humanity's place in the universe. In the following section, I examine the use of anthropomorphic metaphor in science popularization generally and in popularizations of quantum mechanics specifically, indicating contexts in which this metaphor is productive and useful, and other contexts in which it is unstable and problematic.

Anthropomorphism, Anthropocentrism and "Quantum Consciousness"

In his introduction to *The Faber Book of Science*, John Carey asserts that anthropomorphism (he uses the term "animism") is "a persistent ally in the popular science-writer's struggle to engage the reader's understanding," and notes its appearance in many of the pieces in his collection (xviii–xix). For Carey, this kind of anthropomorphism is unavoidable: "All science is inevitably drenched in our human presumptions, designs and concepts … . From this viewpoint, to say that iron 'breathes' is no more absurd than to say that it is called 'iron', or that its chemical symbol is Fe" (xviii). However, recognizing the anthropomorphic nature of all language, as with recognizing the metaphorical nature of all language, should not foreclose attempts to identify contexts in which anthropomorphism works in especially slippery (sometimes strategically slippery) ways.

One literary critic who has extensively examined anthropomorphism in scientific writing (particularly Darwin's) is Beer. While acknowledging, like Carey, the anthropocentric nature of language itself, Beer insists on attention to context: "not all potential significations are active. One of the most remarkable powers of the human mind—less often commented on than its power to proliferate senses—is its power to exclude, or suppress, feasible meaning" (*Open Fields* 156). As outlined above, for Beer the context in which a metaphor is deployed is central to determining how it is likely to be read, that is, "how far metaphors may overturn the bounds of meaning assigned to them" (*Darwin's Plots* 50). The "potential significations" of anthropomorphic metaphor are more likely to be activated in an argument that has obvious consequences for humanity's place in the world, such as that presented in *The Origin of Species*. Thus Beer asks to what extent Darwin's use of anthropomorphism is unavoidable (due to the nature of language), and to what extent it is knowing and strategic (*Darwin's Plots* 57). These questions are particularly relevant to a text such as Darwin's, which reached beyond a specialist readership. As Beer emphasizes, the constraints which limit the interpretations of certain words within a scientific readership disappear in a heterogeneous, general readership (47, 49). Hence, while Darwin assumed that his readers would automatically recognize the term "natural selection" as a convenient metaphorical shorthand, this was not in fact the case. Beer explains, "The problem, of course, was that every one did *not* know what was meant by natural selection—the term was a neologism and

---

animal behaviour. Crocker argues that popularizations make plain the "anthropomorphic impulses" that professional scientific language "represses" (162).

therefore stood forth with full metaphorical expressiveness and personifying power" (*Darwin's Plots* 63). Some of Darwin's readers focused on the element of intention implied by the metaphor, and then literalized the metaphor by assuming that organisms have conscious choice in the way they are modified (62–3). Beer notes that in the second edition of *The Origin of Species* Darwin adds the phrase "It may be metaphorically said" to one particularly ambiguous passage (63), presumably to prevent this kind of overly literal reading. Metaphors are always open to new interpretations, and meaning is kept in check only by contextual limits. Scientists are themselves well aware of this: like Darwin, they often draw the reader's attention to metaphorical uses of language by adding inverted commas or explicit remarks. Erwin Schrödinger, after deploying an anthropomorphic metaphor in his *Mind and Matter*, writes "I wish to underline three times in red ink that I mean this only as metaphor" (103).

Popularizers of quantum theory have been employing anthropomorphic metaphor since the theory was in its early stages: "An electron is like a man," wrote Bertrand Russell in his popularization *ABC of Atoms*, published in 1923, "who, when he is insulted, listens at first apparently unmoved, and then suddenly hits out" (63). In some texts, an extended anthropomorphic technique provides the entire expository framework. In these books, which might be termed "quantum fables," exposition is achieved by endowing subatomic entities with elaborate human characteristics and behaviour; there is an exchange between macroscopic and microscopic worlds, and a human is able to interact with the subatomic environment. The best-known examples of this form are George Gamow's "Mr Tompkins" essays on physics. Employing the dream-sequence, Gamow inserts his mild-mannered bank clerk Mr Tompkins into realms of exaggerated scale, where quantum effects act on human-sized objects.[4] In one episode, "The Gay Tribe of Electrons" (*Mr Tompkins in Paperback* 112–27), Mr Tompkins actually becomes an electron, and is pictured orbiting an atomic nucleus, talking and physically interacting with other electrons, which are themselves presented as "vague, misty" humanoid forms (114). A more recent and self-conscious example of the "quantum fable" approach can be found in Robert Gilmore's *Alice in Quantumland: An Allegory of Quantum Physics* (1994). The crudest in a long line of popularizations which borrow from Lewis Carroll,[5] Gilmore's heavily illustrated text begins with a bored Alice becoming absorbed by the coloured pixels of her television set, tripping over her copy of "'Alice in Wonderland'" (1), and finding herself in a world inhabited by quantum entities of her own size. Carroll's familiar figures make their appearances, suitably transformed: the white rabbit who can't find his keys quantum tunnels through his door (8–9); Schrödinger's—rather than the Cheshire—Cat appears in the bough of a tree and then vanishes (55); Tweedledum and Tweedledee become the two detectors in the

---

4 Gamow also used the form of the fable to popularize a number of other scientific fields, including relativity (in *Mr Tompkins in Wonderland*); in this case, the exaggeration of scale occurs in the opposite direction, and relativistic effects become noticeable in everyday human activities.

5 See Sigman (28–9) for a brief discussion of popularizations that allude to Carroll's work.

EPR paradox (191). Russell Stannard's popularization for children *Uncle Albert and the Quantum Quest*, published in the same year as *Alice in Quantumland*, employs a near-identical framework. However, Stannard adds another metaphorical layer to his text by naming his "Alice" character "Gedanken," and having her enter her adventure through a thought bubble generated by her physicist uncle, Albert.[6] When Gedanken returns, she reports her findings to Uncle Albert, and together they formulate the laws of quantum physics. Thus Stannard metaphorically conveys a particular tool for developing scientific theory (the thought experiment) as well as the theories themselves, and neatly avoids the use of explanatory "boxes" which children might ignore.

Other popularizations use miniature versions of the "quantum fable" in isolation to explain scientific concepts. Gribbin, in his *In Search of Schrödinger's Cat*, compares the behaviour of fermions and bosons to that of the respective audiences at two events he attended, a play starring Spike Milligan and a Bruce Springsteen concert. The theatre audience, on Milligan's request, left their assigned seats and filled up the empty seats close to the stage, "acting like nice, well-behaved fermions, each person occupying just one seat (one quantum state) and filling up the seats from the most desirable 'ground state,' by the stage, outward"; at the Springsteen concert, by contrast, the audience rushed up and squashed against the stage: "All of the 'particles' crammed into the same 'energy state' indistinguishably—and that is the difference between fermions and bosons. Fermions obey the exclusion principle, bosons do not" (98). Likewise Danah Zohar in *The Quantum Self* explains the notion of virtual transitions by an extended analogy with a "quantum hussy" approached by several suitors: "in the quantum world, the dizzy girl would simply take up with *all* suitors, *all at once*, perhaps even setting up house with each of them simultaneously" (16). Although eventually she would have to "settle down, marry and live in one house with just one of the suitors," she would nonetheless have left "'traces' of herself in the various neighbourhoods where she had occupied temporary addresses"— possible offspring, and so forth (16). In each case, the popularizer uses the analogy as a chance to introduce some extraneous "human interest." For Zohar's purposes, any example in which a person pursues two simultaneous but macroscopically contradictory courses would suffice, yet she revels in her description of the scenario. The girl is initially described as a "sheltered young woman" who becomes "overexcited" upon her "'coming out'"; her parents might be "scandalized" but would have to write to all her addresses to reprimand her. Gribbin's metaphor also contains peripheral detail—what Milligan said, which song Springsteen opened with. Both analogies could potentially, in Beer's phrase, "overturn the bounds of meaning assigned to them": Zohar's readers might choose to focus on the immoral rather than the transitory connotations of the promiscuous "quantum hussy," and view the quantum world as a challenge to traditional mores; Gribbin's might fix on civility, rather than orderliness, as the quality which should be attached to fermions, and view them as somehow superior to boisterous bosons. But the likelihood of

---

6   *Uncle Albert and the Quantum Quest* is the third in a series of "Uncle Albert" books by Stannard, all of which employ the "Gedanken" character and the "thought bubble" metaphor. The first two books are popularizations of special and general relativity.

such interpretations is low: neither popularization provides a context that would nourish and sustain them, and the explicitness of the comparisons in each case reduces what Black might term their "suggestiveness."

In these two examples, as well as in the extended "quantum fables" described above, the use of anthropomorphism is well acknowledged and elaborately constructed. More typically, popularizers will employ particular verbs that impart human characteristics to non-human phenomenon. For example, the word "know" is often employed as a metaphorical shorthand to describe seemingly inexplicable interactions between an object and its environment. Paul Davies uses this device in his discussion of the double slit experiment: "How does any electron know what the other electrons, maybe in other parts of the world, are going to do?" (*Other Worlds* 66). Gribbin uses similar language: "The electrons not only know whether or not both holes are open, they know whether or not we are watching them, and adjust their behaviour accordingly" (*In Search of Schrödinger's Cat* 171). Richard Feynman asks how a photon can "'make up its mind' whether it should go to A or B" (*QED* 18). Habitual anthropomorphism of this kind can be the source of considerable confusion. The presentation of an electron as a decision-maker, for example, is problematic in that it contradicts one of quantum physics's most basic and radical pronouncements: that the behaviour of individual subatomic entities is totally random. If one wants to relieve oneself of the burden of a decision, one resorts to an effectively random process, such as a coin toss; yet popularizers continually project knowledge and volition onto the fundamental form of randomness that characterizes the behaviour of quantum entities. This leads to some oxymoronic pronouncements: veteran popularizer Martin Gardner, in a discussion of the misuse of quantum physics by "fringe" scientists such as parapsychologists, declares that "nature *decides by chance* whether to give the particle a plus or minus spin" (*The New Age* 109; emphasis added).

This example indicates the kind of difficulties which anthropomorphic metaphor can produce when applied to quantum phenomena; nevertheless, in the examples from Feynman, Davies and Gribbin quoted above, the anthropomorphic metaphors deployed are relatively stable. Although phrases such as "the electron knows" or "makes up its mind" or "adjusts its behaviour" could potentially impart a sense of consciousness to the electron (just as the phrase "natural selection" for some of Darwin's readers implied a deliberate choice), the context of the metaphor in each case does not encourage such an interpretation. To use Beer's terms again, the "potential significations" of this metaphor are unlikely to be "activated." These metaphors in fact rely for their effectiveness on the reader's recognition that they do *not* suggest that electrons are conscious. When popularizers ask how an electron "knows," assuming their readers believe that electrons *cannot* "know" as humans do, they are emphasizing that the interaction in question is thus inexplicable, at least in terms of previous understandings of interactions. They are then in the position to introduce or emphasize the radical (although still metaphorical) notion that a particle can act like a wave, spread out in space. It is context—the service to which a metaphor is put, the wider argument in which it is embedded—that determines the range of interpretations it is likely to generate.

Within popularizations of quantum theory, a context that *is* likely to activate the "potential signification" of anthropomorphic metaphor is one in which the consequences of the theory for humanity's image of itself is brought to the fore. There is a clear message in a number of popularizations that quantum physics, with its recognition of the embeddedness of the observer, gives a new status to human consciousness, and thus restores humanity to centre stage after four hundred years of post-Copernican exile. In these popularizations, a resonance is created between the perceived anthropocentric elements in quantum theory, and the anthropomorphic metaphors employed to explain the theory. The result is that the ambiguity of these metaphors is amplified by potential philosophical implications, rather than damped by scientific constraint. It is to these texts that we need to apply the kinds of questions that Beer applies to *The Origin of Species*: to what degree do the anthropomorphic metaphors employed in these texts exceed their ostensible function? What human characteristics do the readers of these popularizations attach, perhaps subconsciously, to quantum entities, and what is the consequent effect on their understanding of quantum theory? Before these questions can be addressed, however, it is necessary to discuss briefly the role of consciousness within quantum theory and its popularization.

Although the radical re-evaluation of the observer-observed relationship implied by Heisenberg's Uncertainty Principle is well known, the exact nature of consciousness in the quantum mechanical measurement process has been debated by physicists and philosophers for decades. In the Copenhagen interpretation, a quantum mechanical system remains in a superposition of a number of possible states, the probabilities of which are described by a "wave function," until a measurement is taken, at which point the wave function "collapses" into one of these states. This collapse is triggered by the interaction of the microscopic system with macroscopic measuring instruments. "The conscious observer," popularizer John Polkinghorne comments, "can then take note if he wishes to do so"; the observer is integral to the system only in that s/he chooses "the disposition" of the classical measuring instruments (*The Quantum World* 63, 78).

There are obvious difficulties, however, in distinguishing between the microscopic system being measured and the macroscopic measuring system. As Davies notes, "even Bohr conceded that [the measuring apparatus] must be subject to the minute uncertainties that are the characteristic of quantum physics. There is no clear dividing line between what is a microscopic system and what is a macroscopic measuring device" (*Other Worlds* 128). This "measurement problem" is highlighted by thought experiments such as the "Schrödinger's Cat" and "Wigner's Friend" paradoxes, which entail sentient creatures existing in a superposition of contradictory live/dead states until they are externally observed. To address these difficulties, Eugene Wigner and others suggested an interpretation of quantum theory that postulates the conscious observer in the place of the classical measuring equipment as the mechanism for the wave function collapse. Some speculative theories, such as John Wheeler's "participatory universe," go so far as to claim that the reality of past events such as the Big Bang relies on the eventual appearance of conscious observers.

The idea that consciousness has some integral role in the physical world has been dwelled upon by popularizers of quantum theory since its inception. Eddington, towards the end of *The Nature of the Physical World*, lists the changed understanding of consciousness due to the new physics as one of his central philosophical points: "Recognising that the physical world is entirely abstract and without 'actuality' apart from its linkage with consciousness, we restore consciousness to the fundamental position instead of representing it as an inessential complication ..." (317). In the post-Capra period, "quantum consciousness" (for want of a better term) has become the central theme of numerous popularizations.[7] As Victor Stenger observes in *The Unconscious Quantum*, "The popular literature abounds with this theme as paranormalists of every stripe, from psychics to astrologers to physicists to cosmologists, proclaim the oneness of the human mind and the fabric of the cosmos" (270).

The philosophical implications of quantum theory are not straightforward, however, and idealist interpretations such as that proposed by Eddington are by no means universally supported. "While there is little doubt that at the operational level [quantum] theory is a brilliant success," writes Davies, "... nevertheless the epistemological and metaphysical aspects of quantum physics continue to cause anxiety" (*Other Worlds* 126). Many physicists believe there is an over-emphasis on consciousness in some philosophical interpretations of the theory. Polkinghorne states that "there is something very unattractive about this particular suggestion [that consciousness triggers the wave function collapse]. It is astonishingly anthropocentric ..." (*Quantum World* 66). Stenger likewise attacks the anthropocentrism of "the myth of quantum consciousness," arguing that it taps into the human unwillingness to accept post-Copernican peripherality (286, 17–19). John Bell (the creator of "Bell's Theorem," a mathematical reformulation of the EPR paradox often used to support New Age speculations) has similarly identified "a myth ... that quantum theory has undone somehow the Copernican revolution." He writes that although "from some popular presentations the general public could get the impression that the very existence of the cosmos depends on our being here to observe the observables," he sees "no evidence that it is so in the success of contemporary quantum theory." He concludes that "it is not right to tell the public that a central role for consciousness is integrated into modern atomic physics" (170).

My point in the above discussion is not to suggest that all attempts to connect quantum physics and consciousness are fruitless or misguided, or to evaluate such attempts in any way, but to contextualize them by indicating their contentious

---

7   Examples include Wolf, *Mind and the New Physics*; Penrose, *The Emperor's New Mind* and *Shadows of the Mind*; Squires's *Conscious Mind in the Physical World*; Zohar; Goswami; Herbert, *Elemental Mind*; and Stapp. These books vary widely in their approaches and claims; certainly not all of them could be considered to adopt a "New Age" or semi-mystical approach. However, they give an indication of the amount of public interest in "quantum consciousness" in the late twentieth century. More recently, the film *What the Bleep Do We Know!?* (2004), a physics popularization of sorts, explored the relationship between quantum theory and consciousness by juxtaposing a fictional narrative with a series of interviews with commentators (including Wolf and Goswami).

position within physics. The rest of this chapter gives a close analysis of *The Dancing Wu Li Masters*, a popularization that does make claims (although not always explicitly) for a relationship between human consciousness and quantum theory. In this analysis, I am not so much concerned with the rights or wrongs of such claims as the way in which they can be introduced figuratively rather than literally. In other words, I want to trace the process through which anthropomorphic metaphor becomes a strategic tool in Zukav's argument.

Slippery Metaphors in *The Dancing Wu Li Masters*

The following analysis points out some problems that Zukav's text, if absorbed uncritically, can generate due to the ambiguity of its metaphors. The central role Zukav's book has played in disseminating ideas from physics to non-scientists make such analysis necessary and important. Despite its unconventional title, *The Dancing Wu Li Masters* is not a peripheral text within the genre of popular physics books. It was a bestseller when first published, and (if Amazon rankings are an appropriate indication) continues to outsell more "mainstream" and more recent bestsellers such as Gribbin's *In Search of Schrödinger's Cat*. Leon Lederman in his popularization *The God Particle* complains that the reading public is too heavily influenced by quantum-mysticism books, and identifies *The Dancing Wu Li Masters*, along with *The Tao of Physics*, as the most prominent (190). Lederman does not dismiss these books wholesale, noting that both authors "have gotten a lot of things right" and produced "some good physics writing"; but he criticizes their tendency to jump from effective explanation to "bizarre" interpretation (190). He singles out for criticism Zukav's apparent ascription of consciousness to photons: "This is fun, perhaps even philosophical, but we have departed from science" (191). There are many relatively obscure physics popularizations that are far more speculative; what makes *The Dancing Wu Li Masters* significant is its reach and impact. As evidence of Zukav's influence, Lederman recounts a conversation he had with a US Senator whose support was important to the funding of a particle accelerator: the Senator, to Lederman's consternation, is interested only in talking about *The Dancing Wu Li Masters* (189).

Equally significantly for the purposes of this discussion, Zukav's popularization has been particularly influential on literary critics. A rough indication of which popularizers literary critics turn to in order to learn about quantum theory can be gathered from an examination of relevant information in the Arts and Humanities Citation Index. I focused on contemporary (post-1975) popularizations, as these are the texts with which this book is primarily concerned. My initial reading of literary criticism dealing with quantum theory identified six commonly cited contemporary popularizations, Capra's *The Tao of Physics* (1975), Zukav's *The Dancing Wu Li Masters* (1979), Davies's *Other Worlds* (1980), Pagels's *The Cosmic Code* (1982), Gribbin's *In Search of Schrödinger's Cat* (1984), and Herbert's *Quantum Reality* (1985). With the exception of Gribbin's text, all of these also appear on Carson's list of "the scientific sources that get cited" by critics interested in quantum theory (644). I investigated these six popularizations using the Arts and Humanities Citation Index (indexing begins in 1986, when all of these texts were available).[8] This revealed that the two New Age popularizers, Capra and Zukav, have the highest "impact factor": a total of 17 articles cited *Other Worlds*, 18 cited *The Cosmic Code*, 35 articles cited *In Search of Schrödinger's Cat*, 42 cited *Quantum Reality*, 93 articles cited *The Dancing Wu Li Masters* and 158 articles cited *The Tao of Physics*. Although Capra was cited more often than Zukav overall, when the search was limited to publications in literary journals, Zukav was the more popular choice. One literary critic states that "any number of popular accounts have been published to make those implications [of quantum theory and relativity] accessible to the non-physicist," but remarks, "Unfortunately, these accounts are of varying quality. A good starting point is Gary Zukav, *The Dancing Wu Li Masters: An Overview of the New Physics* ..." (Booker 577, 585). This advice would be readily understandable in the early 1980s, when sources of quantum theory accessible to the layperson were few. However, this critic is writing in 1990, and yet recommends Zukav's text over well-known "mainstream" popularizations, such as those by Davies, Pagels and Gribbin. Yet, despite the prominence of Zukav's text within the popular physics genre generally and in its impact on "literature and science" criticism, it has never itself been examined from a literary perspective. While *The Tao of Physics* has been the subject of close textual analysis (Restivo, "Parallels and Paradoxes I"), Zukav's popularization seems always to be treated as a pure information source, never a textual construction, by critics in the humanities.

Why, one might ask, do literary critics favour a text with twelve "Chapter 1"s,[9] and which, according to one contemporary reviewer, "sounds as if it had been written for people under the influence of soft drugs" (Sokolov)? A possible explanation is a perceived connection between the New Age and literary critical embracements of quantum theory. One critic who recommends Capra's text suggests that it is particularly suitable because "it approaches physics as a language and compares modern physics to the teaching literatures of Japan and China—a more familiar ground for most literature students" (Kemnitz 58 n. 4). Another reason could be the fact that Zukav (unlike Gribbin, Davies and Capra) has no training in physics, and makes a point of constructing a shared identity with his non-scientific readers. Using an implicit "two cultures" model of disciplinary relations, Zukav argues that people fall into "two categories of intellectual preference"—they have either a "scientific mental set" or a "liberal arts mental set" (23)—and he aligns himself and his reader with the latter: "I hope this book will be a useful translation which will help those people who do not have a scientific mental set (like me) to understand the extraordinary process which is occurring in theoretical physics .... For better or worse, my first qualification as a translator is that, like you, I am not a physicist" (24). Zukav states in his introduction that his book is about "quantum physics and relativity," not "physics and eastern philosophies," although the "poetic framework"

---

8   I conducted this investigation via the "ISI Web of Knowledge" database in late 2005.

9   This numbering is presumably in accordance with the philosophy of one of Zukav's sources of inspiration, T'ai Chi Master Al Huang: "'Every lesson is the first lesson'" (Zukav 35).

he employs is "conducive to such comparisons" (25). He thus aligns both himself and his approach with literary rather than scientific modes of understanding.

Zukav's unusual title itself indicates that its author is more consciously interested in word-play than most popularizers. The phrase "Wu Li" is notable for the multiple meanings it can generate simultaneously. "Wu Li," Zukav states, means "Patterns of Organic Energy," and is used as term for physics in Taiwan (31). Emphasizing the non-phonetic nature of Chinese ideograms, and pointing out the dependence of the Chinese language on syllabic inflection, Zukav explains that the phrase written "Wu Li" in English can have multiple interpretations in Chinese, and he lists four additional meanings: "My Way," "Nonsense," "I Clutch My Ideas" and "Enlightenment" (33). Each of these five interpretations becomes the title and thematic centre of the five sections of Zukav's text. Moreover, the egocentrism implied by the titles ("I"/"My") plays an important part in Zukav's presentation of quantum physics.

Zukav's title also introduces the anthropomorphism that plays such a significant role in the text, through the extended metaphor of "the dance."[10] On one level the phrase "Dancing Wu Li Masters" refers to physicists, or rather a particular group of physicists who, unlike mere "technicians," seek "the true nature of physical reality," deal with "the unknown," and have reached a degree of enlightenment about their subject comparable to that of an Eastern Master (36, 42). Einstein, we are told, "was perhaps a Wu Li Master." Zukav introduces the dance metaphor by stating that a Wu Li Master "dances with his student," and he (in Zukav's text all Masters are male) knows that he is not explaining the world but rather "dancing" with it (35). From this point on, this metaphor appears variously and frequently. It is often used, in a sense reminiscent of Yeats,[11] to indicate that participators are inextricably part of and defined by the system in which they participate: "The Wu Li Masters move in the midst of all this, now dancing this way, now that. ... Now they become the dance, now the dance becomes them" (43); "At the subatomic level the dancer and the dance are one" (212). But Zukav also employs it more specifically to describe the movements of physical entities: "Waves are playful creatures that like to do dances of their own" (82); "Subatomic particles forever partake of this unceasing dance of annihilation and creation. In fact, subatomic particles *are* this unceasing dance of annihilation and creation" (235). Thus, as the text progresses,

---

10 Zukav's application of the dance metaphor to modern particle physics is not new: four years before *The Dancing Wu Li Masters* was published, Capra had extensively discussed the Eastern origin of this metaphor and applied it directly to quantum theory. A major theme of *The Tao of Physics* is the way in which the interaction of particles can be represented as a ceaseless dance, which parallels the Dance of Shiva, the Hindu Lord of the Dancers (preface to 1st edn, ch. 15). Zukav mentions Capra's text only in two unindexed footnote references, neither of which refers to the dance concept. While both popularizers link the dance metaphor to Eastern cultural traditions, it also has a long tradition in the West. According to Alan Brissenden, not only is the notion of a cosmic dance an ancient one in Western culture, but also the concept of a *dance of atoms* has been established since at least the second century (3).

11 "O body swayed to music, O brightening glance, / How can we know the dancer from the dance?" ("Among School Children," verse 8, lines 63–4).

the dance becomes a metaphor for the interaction of subatomic particles, and in a wider sense, all matter/energy, as well as the relationship between physicists and their research. The anthropomorphism evident here, as so often in Zukav's text, is of the type which Beer might term "strategic" rather than unavoidable. The dance metaphor establishes an implicit connection between quantum phenomena and consciousness (physicists' intellectual processes). This connection is reinforced by further metaphors and loaded adjectives: the philosophical implications of quantum mechanics are termed "psychedelic" (35); "imagination" is described as "physics come alive" (35); quantum leaps are described as "idea-like characteristics" because ideas also "can and do change discontinuously" (104). Thus a network of interlocking metaphors, all of which identify quantum theory with the human mind, is gradually built up.

Although the association between human consciousness and subatomic particles is established through this dance metaphor, on a literal level Zukav tends to downplay the importance of this theme. One early hint is provided when Zukav mentions that the person who invited him to the "conference on physics" which inspired the book was the "director of the Physics/Consciousness Research Group" (31). However, the reader must search fairly carefully to realize that the conference appears to have been more specifically a "Conference on Physics and Consciousness" (337, n. 5). It also requires attention to detail as well as a wider knowledge of the genre to recognize that at least four of the five physicists who "read and commented upon the entire manuscript" are themselves interested in physics-and-consciousness issues.[12] Zukav provides only one page reference (p. 102) under the index entry "Consciousness, physics and," despite the fact that the word "consciousness" itself appears in a highly relevant context on at least six other occasions (56, 88, 107, 240, 323, 326), and general discussion of the connection between "the mind" and quantum reality occurs frequently. This lack of acknowledgement seems less strange when one realizes that Zukav's argument relies on the reader absorbing claims about quantum physics and consciousness without being presented with any explicit argument for these claims: the argument instead is embedded in Zukav's anthropomorphic metaphors and analogies. To challenge, or even to identify Zukav's claims, the reader must turn to literary criticism rather than scientific evidence.

Zukav first introduces ideas about consciousness in his second chapter, via a technique which one critic, Sal Restivo, has labelled "the parallelist method of juxtaposed quotations" (Restivo, "Parallels and Paradoxes I" 148):

> The new physics, quantum mechanics, tells us clearly that it is not possible to observe reality without changing it .... .

---

12 Zukav lists these names in his acknowledgments (7). Jack Sarfatti is the director of a Physics/Consciousness Research Group mentioned above. Henry Stapp has written a popularization dealing with quantum physics and consciousness. David Finkelstein attended the same conference on physics and consciousness as Zukav and Sarfatti. Brian Josephson, along with V. S. Ramachandran, has edited the proceedings of an interdisciplinary symposium on consciousness (University of Cambridge, January 1978) entitled *Consciousness and the Physical World*, and is interested in both Eastern philosophy and psychic phenomena.

> According to quantum mechanics there is no such thing as objectivity. We cannot eliminate ourselves from the picture. We are a part of nature, and when we study nature there is no way around the fact that nature studies itself. Physics has become a branch of psychology, or perhaps the other way round.
> Carl Jung, the Swiss psychologist, wrote:
> The psychological rule says that when an inner situation is not made conscious, it happens outside, as fate. That is to say, when the individual remains undivided and does not become conscious of his inner contradictions, the world must perforce act out the conflict and be torn into opposite halves.
> Jung's friend, the Nobel Prize-winning physicist Wolfgang Pauli, put it this way:
> From an inner center the psyche seems to move outward, in the sense of an extraversion, into the physical world ...
> If these men are correct, then physics is the study of the structure of consciousness. (Zukav 56).

This passage begins in a manner fairly typical of quantum physics popularizations, with Zukav emphasizing the inevitable effect of the experimental system on its object of study. However, he ventures into more speculative territory when he vaults over the difficulties involved in moving from the effect of observation on the observed system to the effect of *consciousness* on the system, and then from consciousness to psychology. He concludes his argument by citing two authorities from opposing "sides" of the two cultures divide, Jung and Pauli. Restivo has analysed this "parallelist method" in his discussion of Capra's *The Tao of Physics*. He argues that this method, which rests on the "basic assumption" that "if the rhetorical, imagery, and metaphoric content of statements on physics and mysticism is similar, the conceptual content must be similar" ("Parallels and Paradoxes I" 151), is marred by a number of general difficulties (153–5). In Zukav's example, it is not clear how either quotation relates to quantum physics and the role of the observer in experimental systems, nor is the context of either statement made clear;[13] their content hardly seems to justify his bold conclusion.

In the third chapter (that is, the first "Chapter 1" of the second "Part 1"), Zukav effectively establishes consciousness as a central theme of his text. The basic rhetorical structure of his argument centres on the term "Wu Li." As mentioned above, one of the possible interpretations of "Wu Li" is "Patterns of Organic Energy," and Zukav employs this metaphor as the title of his second "Part 1." The first chapter of this section is entitled "Living?" At the seminal points of this chapter, Zukav relies very little on scientific evidence or authority, but rather persuades by means of anthropomorphic analogy and metaphor. His initial strategy

---

13 Zukav's endnote reveals that the quotation from Pauli derives from his essay "The Influence of Archetypal Ideas on the Scientific Theories of Kepler," which was published together with an essay by Jung in their *Interpretation of Nature and the Psyche* (Pauli's quotation is on p. 175). Pauli does not mention quantum physics in connection with this statement, and it is difficult to see how, in context, it implies that "physics is the study of the structure of consciousness." When Pauli does turn to quantum physics towards the end of his essay, it is to observe that the complementarity principle suggests that the physical and the psychical can be seen as two complementary sides of reality, as are the wave and the particle (208). This is significantly different from the conclusion Zukav draws.

in this chapter is to provide his own idiosyncratic interpretation of the relationship between "organic energy" and quantum physics:

> When we talk of physics as patterns of organic energy, the word that catches our attention is "organic." Organic means living. Most people think that physics is about things that are not living, such as pendulums and billiard balls. This is a common point of view, even among physicists, but it is not as evident as it may seem. (70)

Zukav has earlier introduced his reader to the New Age buzz-word "organic," noting that the participants at the physics conference he attended at Big Sur ate "organic food" (31). In the paragraph above, his definition—"Organic means living"—is a simplification of the usage favoured by New Age devotees and popularizers, for whom the term usually implies a dynamic unity like that found in living beings. Capra, for example, states, "In contrast to the mechanistic Western view, the Eastern view of the world is 'organic'.... The cosmos is seen as one inseparable reality—forever in motion, alive, organic; spiritual and material at the same time" (29). Zukav goes on to develop his definition via an imaginary dialogue[14] conducted with a hypothetical person Jim de Wit, "the perpetual champion of the non-obvious" (70). Jim's argument is anthropomorphic, and I will briefly summarize the gist of the dialogue (70–72; original emphasis), as it is an integral part of Zukav's textual strategy.

Physics applies to living things as much as to non-living things, argues Jim—in a vacuum, humans accelerate at the same rate as rocks. But unlike rocks, "we" (layman Zukav and his reader) retort, humans exercise choice: "humans process *information* (they know they may be hurt) and they *respond* to it (by not falling)." Jim replies that it may only be our perceptual limitations that prevent us from observing these characteristics in other entities; plants are seen to respond to stimuli when we watch them using time-lapse photography: "'If this is so, then how can we say with certainty that rocks, and even mountain ranges, do not react also as living organisms?'" Jim goes on to claim that inert matter responds to stimuli: "'Under the right conditions, for example, sodium reacts to chlorine (by forming sodium chloride—salt), iron reacts to oxygen (by forming iron oxides—rust), and so on, just as humans react to food when they are hungry and affection when they are lonely.'" When "we" reply that "'it hardly seems fair to compare a chemical reaction to a human reaction,'" Jim counters that we do not know that "'our responses are not as rigidly preprogrammed as those of a chemical, with the only difference being that our programs are enormously more complex.'" Thus if we do not consider inanimate matter such as stones to be organic or "living," we cannot consider ourselves organic, and thus cannot consider ourselves alive. "Since this is absurd," concludes Zukav, "the only alternative is to admit that 'inanimate' objects may be living." Drawing on this argument, Zukav asserts that:

> The distinction between organic and inorganic is a conceptual prejudice. It becomes even harder to maintain as we advance into quantum mechanics. Something is organic,

---

14 The imaginary dialogue is a literary form with a long tradition in science popularization. See Myers, "Science for Women" and "Fictions for Facts."

according to our definition, if it can respond to processed information. The astounding discovery awaiting newcomers to physics is that the evidence gathered in the development of quantum mechanics indicates that subatomic "particles" constantly appear to be making decisions! (72)

No doubt many cognitive scientists would take issue with Zukav's extended definition of "organic" (in other words, "living") as something which "can respond to processed information"—by this definition, an electronic auto-bank is certainly a living creature; so is every computer, in fact. But even if we accept this definition, Zukav's argument is self-undermining. Beneath the rhetoric of the dialogue is an implicit reductionist argument: if human (organic) reactions are just a more complex version of chemical reactions, there can be no difference between the properties of the two, and chemicals must therefore be organic (in Zukav's sense) also.[15] The reductionism involved in terming subatomic entities "organic" is also at work in Zukav's blatantly anthropomorphic claim that these entities "constantly appear to be making decisions": a quality which belongs to complex whole (a human) is endowed upon the subatomic parts of this whole.[16] In *The Dancing Wu Li Masters* this reductionist form of argument is highly ironic, given the holistic conclusion to which it leads: "the philosophical implication of quantum mechanics is that all of the things in our universe (including us) that appear to exist independently are actually parts of one all-encompassing organic pattern, and that no parts of that pattern are ever really separate from it or from each other" (Zukav 72–3).

Admittedly, Zukav does not state in the above quotation that subatomic entities *are* "making decisions," just that they *appear* to be; and occasionally he identifies the word "know" as a metaphor by giving it quotation marks. These markers of metaphor diminish, however, as the text develops. The qualifying word "appear" eventually *dis*appears, and the anthropomorphic logic is reinforced by the use of the word "know":

> ... subatomic particles constantly appear to be making decisions! More than that, the decisions they seem to make are based on decisions made elsewhere. Subatomic particles seem to know *instantaneously* what decisions are made elsewhere, and elsewhere can be as far away as another galaxy! The key word is *instantaneously*. How can a subatomic particle over here know what decision another particle over there has made *at the same time the particle over there makes it?* All evidence belies the fact that quantum particles are actually particles. ... [Instantaneous communication] means that "particles" may not be particles at all. (72)

Zukav seems poised here to introduce the wave-particle duality; and, indeed, a few paragraphs later he begins to trace the evolution of this idea: "To understand these decisions and what makes them, let us start with a discovery made in 1900 by Max Planck ..." (73). If we were to interpret literally Zukav's earlier claim that particles make, or appear to make, decisions, the answer to "what makes them?"

---

15 This is a critique that Ken Wilber in his *Quantum Questions* applies more generally to attempts to reconcile physics and mysticism (27–8).

16 See Rose, Lewontin, and Kamin (278) for an analysis of the way a similarly reductionist reasoning leads to genes being labelled "selfish" or "homosexual."

would simply be "the particles"; thus, this sentence suggests that the first claim was metaphorical. Zukav is set to follow the usual expository pattern, in which questions about electrons' apparent "knowledge" and "decision-making" are addressed by introducing a completely new framework. However, although the remaining few pages of the chapter highlight the bizarre nature of the wave-particle duality, and introduce the idea of "probability waves," Zukav does not go on to emphasize the way in which a "wave" formulation might shed light on apparent decision-making, at least not to the degree that would be required to counteract the force of his previous anthropomorphic rhetoric. Instead, he encourages the concretization of his original metaphor by stating excitedly that "Some physicists ... speculate that photons may be *conscious*!" (88). The scientific evidence cited to support this is slim—one quotation from E. H. Walker, who is the leading theoretician of "paraphysics," the attempt to use physics to support parapsychology (M. Gardner, *The New Age* 111), although Zukav fails to inform his reader of this. However, to the lay reader the idea of "conscious" subatomic entities may not seem in need of much supporting evidence, as it is an assumption that has been implicit in Zukav's anthropomorphic language since the beginning of the chapter. Similarly, when he comes in the next chapter to discuss the more common question of the role of consciousness in the wave-function collapse, this issue has already been confused by his metaphorical endowment of consciousness upon subatomic particles themselves.

The endowment of agency, and even decision-making, upon subatomic entities is fairly typical of popularizers, even non-speculative ones such as Feynman. These popularizers always run the risk that their metaphors will (to repeat Beer's phrase) "overturn the bounds of meaning assigned to them." However, as noted above, this kind of anthropomorphic metaphor is relatively stable if it has no particular resonance with the broader claims presented in the text. Zukav's use of anthropomorphic metaphor, by contrast, is designed to create such a resonance— to encourage the reader to attribute consciousness to subatomic particles. He has already equated the term "organic" with "living" and suggested that physics might be seen as "the study of the structure of consciousness." Beer has observed the way in which "[s]eemingly stable terms may come gradually to operate as generative metaphors" in *The Origin of Species* (*Darwin's Plots* 50); the same claim might be made of *The Dancing Wu Li Masters*. Words such as "know" and "decision," conventional pedagogical metaphors in many popularizations, take on a dual role as generative metaphors in Zukav's text, as they form a central part of his implicit theory of "quantum consciousness."

The significance of anthropomorphic language is greatly increased when a text deals with anthropocentric notions. How does Zukav's eagerness to endow subatomic entities with human consciousness relate to anthropocentrism? A clue is contained in the summary of his argument provided near the end of the third chapter, a summary which is given impact by its placement just before a break in the text:

> This brings us back to where we started: Something is "organic" if it has the ability to process information and to act accordingly. We have little choice but to acknowledge that photons, which are energy, do appear to process information and to act accordingly,

and that therefore, strange as it may sound, they seem to be organic. Since we are also organic, there is a possibility that by studying photons (and other energy quanta) we may learn something about us. (88)

This summary reveals the degree to which Zukav's overall argument relies on his idiosyncratic use of the word "organic," and hence the rhetoric of his dialogue with Jim de Wit and the passages that follow. Although "decisions" has been lost here—transformed into a more standard anthropomorphism, "act accordingly"—the shadow of the previous metaphor remains. But, more importantly, the summary also indicates the direction in which Zukav's argument points: towards the claim that unlike classical science, which decentres humans by reducing them to the sum of inanimate parts, the "organic" physics of quantum mechanics might resurrect them to new importance.

In the next "Part 1" Zukav develops this anthropocentric argument. It is no coincidence that for the title of this section he chooses the meaning of "Wu Li" which corresponds in English to "My Way," and entitles the first chapter in this section "The Role of 'I.'" He begins the chapter by recounting the beliefs held by humanity "[i]n the days before Copernicus": the belief that the Earth was the centre of the universe; or, in India, the belief that "each person, psychologically speaking ... was recognized as being the center of the universe." He argues that this belief is not egotistical, as it applies to every person. It would be difficult, however, to argue that this is not an anthropocentric position. He goes on to assert that the new physics—relativity and quantum mechanics—is the site of the re-emergence of "an ancient paradigm": "In vague form, we begin to glimpse a conceptual framework in which each of us shares a paternity in the creation of physical reality. Our old self-image as impotent bystander, one who sees but does not affect, is dissolving" (114). Here Zukav employs the first-person plural voice, as he often does, to produce the effect of identification with the reader. Over the next page or so, this identification becomes more significant as Zukav decries the alienating and decentralizing effects of the work of "The Scientists" on "us"—laypeople like the reader and himself, whom he places in direct opposition to "The Scientists." Zukav's language takes on a lyrical quality as he announces humanity's reversal of fortune: "Amid the powerful purr of particle accelerators, the click of computer printouts, and dancing instrument gauges, the old 'science' that has given us so much, including our sense of helplessness before the faceless forces of bigness, is undermining its own foundations." For Zukav, "the faceless forces of bigness" refers not only to the indifferent workings of the post-Copernican universe, but also to the esoteric proceedings of science itself. He goes on to proclaim the good news of centrality regained:

> We have tried to [disown our part in the universe] by relinquishing our authority to the Scientists. To the Scientists we gave the responsibility of probing the mysteries of creation, change, and death. To us we gave the everyday routine of mindless living.
> The Scientists readily assumed their task. We readily assumed ours, which was to play a role of impotence before the ever-increasing complexity of "modern science" and the ever-spreading specialization of modern technology.

Now, after three centuries, the Scientists have returned with their discoveries. They are as perplexed as we are ...

"We are not sure," they tell us, "but we have accumulated evidence which indicates that the key to understanding the universe is *you*."

This is not only different from the way that we have looked at the world for three hundred years, it is *opposite*. (115)

There is a definite note here of triumphant "We told you so!": the resentment of "literary" types (Zukav and his reader) to the Scientists' monopoly on the "big questions" has turned into delight due to this apparent reversal. But this reversal is achieved only through a blatantly anthropocentric presentation of the implications of the embeddedness of the observer in quantum mechanics. And when the kind of anthropomorphic language described above is combined with this anthropocentric philosophy, a space is opened for the reader to embrace what Beer terms the "full metaphorical expressiveness" of the language, and for secondary meanings of metaphors to become literalized and used as an integral feature of this philosophy. Whether or not quantum physics *does* include a central role for consciousness, whether or not photons themselves *are* conscious, is not the issue; the point is that Zukav's exposition of quantum theory rests on the persuasiveness of his anthropomorphism. When lay-readers later come to Zukav's discussion of such concepts as Bell's Theorem and instantaneous action-at-a-distance, which may indeed have implications for our understanding of consciousness and our place in the universe, it would be close to impossible for them not to approach these discussions with a perspective unaffected by his anthropomorphic language.

When the lay-reader is a literary critic, this process is compounded as Zukav's metaphors (which by now may no longer be recognized or presented as such) are placed in new contexts. Thus when Dennis Bohnenkamp, in an article entitled "Post-Einsteinian Physics and Literature," asserts without qualification that "One of the mysterious findings of quantum physics is that matter is not inert or dead, but ... reveals distinct signs of life," it is no surprise to find that one of his sources is *The Dancing Wu Li Masters*. A scientist reading this statement is unlikely to give much credence to the complex parallel that Bohnenkamp draws, in his analysis of James Joyce's fiction, between matter that is somehow "alive" and the epiphanic object as a "sentient participant" (24). There may well be a useful insight here, but the speculativeness and vagueness of the original claim (that matter has "signs of life") renders further analogy so loose as to be worthless. Claims of this sort are thus likely to inflame cross-disciplinary arguments. The critic in question is responsible to some degree for this situation; but equally responsible are the popularizers on whom he draws. Zukav explicitly aims to address the "notable *communication* problem" between people with scientific and literary mindsets (24), but by encouraging readers to take his own "poetic framework" too literally he produces as many problems as he solves.

I am not suggesting here that Zukav's metaphor of decision-making or "knowing" particles is problematic in an absolute sense. This metaphor can be and often is a useful pedagogical tool for popularizers. But in the context of Zukav's overall argument it takes on an extra-pedagogical, strategic role which both encourages and depends upon its literalization. Metaphors may, as Beer argues, exceed their assigned meanings, and in popularizations of quantum mechanics that engage with anthropocentric issues, anthropomorphic metaphors are particularly likely to generate uncontrolled meaning. Some texts, such as Zukav's, actively encourage ambiguity of meaning in order to rhetorically construct an argument for which established scientific evidence is slim. My point here is not that literary critics should not read, or draw upon, Zukav's popularization, but rather that they should not abandon their literary critical apparatus simply because they are dealing with a science book. Before literary critics construct their own metaphors linking scientific and literary concepts, they need to be aware of the metaphors already embedded in the science popularizations they use.

"In talking about the impact of ideas in one field on ideas in another field," observes Richard Feynman in *The Meaning of It All*, "one is always apt to make a fool of oneself" (3). His observation sounds pessimistic until one realizes on reading further that, despite misgivings, he is determined to take on such a task. If there is to be any conversation at all between literature and science, it is important that the exponents of both risk looking foolish by drawing from fields beyond their own. Equally important, however, is that their conversation is as informed and contextualized as possible. As I argued in the previous chapter, the hostilities of the "Science Wars" fed on cross-disciplinary miscommunication and misunderstanding, with each side attacking a distorted image of the other. One means of preventing such distortion is through careful attention to the selection and use of the texts from which information about the unfamiliar discipline is gathered. It is only by looking critically at popularizations, as they look critically at any other sample of writing, that literary critics will be able to gain an informed understanding of the scientific material with which they deal. This does not mean that they should accept only those popularizers, or those statements, approved by the scientific community, and ignore anything speculative, idiosyncratic or unusual.[17] Rather, it means that they should be self-conscious about their choice of popularization; aware that internal differences exist within the scientific community, and that popularizations are not always straightforward translations of agreed-upon fact; and alert to the textual strategies that popularizers employ. Only by paying attention to these strategies will literary critics achieve the degree of finely nuanced argument that will render their analyses meaningful, not only within the community of "science and literature" scholars, but within a broad, multidisciplinary academic sphere. This chapter has highlighted the strategic use of metaphor in a prominent popularization; the next chapter shows how narrative structure can be used in an equally strategic manner.

---

17 There is no guarantee, for example, that popularizations written by scientists are any less ambiguous than those produced by non-scientists. Mara Beller, in her article "The Sokal Hoax: At Whom Are We Laughing?" in *Physics Today*, argues that the popular and semi-popular writings of the founders of quantum mechanics, such as Heisenberg and Bohr, are just as vague and open to misinterpretation as the work of contemporary non-scientist popularizers such as Zukav. She suggests that the real figures of ridicule in the hoax should not have been postmodernist critics, but rather the quantum physicists whose obscure writings provide their support.

CHAPTER 5

Exploding the Big Bang: Popular Cosmology as Mythic Narrative

> In three hours I could break the back of my piece on narrative in science. I already had the outlines of a theory—not one that I believed in necessarily, but I could hang my piece around it. Propose it, evince the evidence, consider the objections, re-assert it in conclusion. A narrative in itself, a little tired perhaps, but it had served a thousand journalists before me. (48)

So muses physicist-turned-popularizer Joe Rose, the protagonist of Ian McEwan's novel *Enduring Love*, until eventually he becomes disillusioned with his own contrived narrative: "What I had written wasn't true. It wasn't written in pursuit of truth, it wasn't science" (50). Rose, who begins by lamenting the demise of the anecdotal form in science, ends by directly opposing his popular narrative to science or "truth."

This chapter looks at the fraught relationship between narrative and science, and the role that popular science books play within this relationship. In particular, it focuses on the much contested border territory between a specific kind of narrative—myth—and a specific scientific theory—Big Bang cosmology. It examines the various competing claims made by different groups on either side of the "two cultures" divide—scientists, popularizers, science studies researchers, social commentators, philosophers of science—about the relationship between scientific and mythic cosmology, highlighting the instabilities of the relationship and the anxieties they generate. I concentrate my close textual analysis on two prominent popularizations of cosmology: Steven Weinberg's *The First Three Minutes: A Modern View of the Origin of the Universe* (1977), one of the bestsellers of the very early years of the popular physics boom; and Stephen Hawking's *A Brief History of Time: From the Big Bang to Black Holes* (1988), the boom's biggest success.[1] I argue that Hawking and Weinberg, while keen to establish a border between scientific and mythic cosmology, nonetheless employ mythic narrative structures in their own popularizations. Constructivist critics are often criticized for equating science with myth; but if the popularizations through which these critics access scientific ideas themselves present science in a mythic framework, the blame must fall equally on both sides of the "Science Wars."

---

1   All citations of Weinberg and Hawking in this chapter refer to *The First Three Minutes* and *A Brief History of Time* respectively, unless otherwise specified.

The Narrative Conventions of Science Writing

One of the ways in which scientific writing has been defined, as I discussed in Chapter 3, is in opposition to literature: as the eschewal, in the words of Thomas Sprat, of "the colours of Rhetorick, the devices of Fancy, or the delightful deceit of Fables" (62). Narrative, which suggests a deliberate arrangement of events in a text by a narrator to achieve a specific effect, signals to McEwan's scientist (in the passage quoted above) a contrivance opposed to the ideal of transparent, objective, descriptive scientific language. Yet specialist scientific writing has, of course, its own sets of narrative conventions, which are arguably just as contrived as any other. Recall Peter Medawar's claim that the scientific paper, with its methods-results-discussion structure, is "a fraud" in the sense that it gives "a totally misleading narrative of the processes of thought that go into the making of scientific discovery" ("Is the Scientific Paper a Fraud?" 233). Its form, he asserts, is designed to imply that "scientific discovery is an inductive process"; it disguises the conceptual biases which scientists bring to their experiments and perpetuates the "fiction" of "innocent observation" (229, 230). Medawar's argument has since been repeated and extended many times by critics within science studies, who point to the suppression in the scientific paper of the messy circumstances of actual research. If an experiment is repeatable, then the details of when, where or by whom it was performed must be unimportant. Similarly, "Irrelevant events, in particular false leads, red herrings and blind alleys, are either excluded or are not tied to subsequent events" (Woolgar, *Science* 77). The remaining narrative thus reflects an idealized application of the "scientific method."

The narrative conventions of popular science writing move away from those of the specialist paper and toward those of the novel. These texts typically reintroduce details of person, place and time that are exiled from the professional report. Particularly in more journalistic popularizations (such as Gleick's *Chaos*), the reader is continuously provided with descriptions of researchers' clothes, hair colour, accent and other personal characteristics. Geographical and historical details erased from scientific reports reappear selectively and often in romanticized form. The narrator may intrude into the text to recall a particularly revealing incident or to voice an opinion. Chapter openings in particular often contain personal images, events and emotions, to draw the reader into the text and create the sense of a "story," in much the same way a novelist might: "I'll never forget the moment I first saw—or, rather, heard—Stephen Hawking. The year was 1969 ..." (Davies, *About Time* 183). Of the opening sentences of the five popularizations that are examined in detail in this book, only one (Waldrop's *Complexity*) focuses directly on the scientific topic it is aimed at expositing; the others are all historical statements involving a human actor.[2]

Popular science, then is caught in a peculiar position: its narratives must retain some sense of the objectivity of science if they are to be accepted as authoritative,

---

2 Here I have ignored prefaces and taken the "opening sentence" of a text to be the first sentence of the author's introduction or prologue, or Chapter 1 where these do not appear.

credible expositions, but they must borrow from the typical style of the novel in order to entertain the wide readership at which they aim. For this reason, critics have noted, many popular science narratives tend to downplay their own literary form (Jurdant 370).[3] The conventions of popular science—such as the insertion of "apparently meaningless detail"—are in this sense similar to those of the realist novel, designed to "pass off whatever degree of fictionality they have as truth" (Kelley 134–5). This means that assumptions (about the world, or about science) embedded in popularizers' narrative structures may not be recognized as such. It is important, then, for critics not to dismiss the narrative conventions of popular science as *only* a kind of formal sugar-coating over the "hard" scientific content, but to subject them to analysis. Ron Curtis, noting the tendency of previous commentators on science writing to "portray narrative simply as a device for engaging the interest of their rather dull readers" (425), claims that the pervasiveness of the narrative mode in popular science goes beyond attention-grabbing: "Popular science, written in a narrative mode, is a powerful tool for promoting a particular normative view of science while, at the same time, rendering that view immune to criticism. It is a way to moralize while appearing only to describe" (434–5).

What morals, then—what "normative view" of science—do popular science books promote? Given the diversity of the popular science genre, there is unlikely to be one answer to this question; critics' role is to identify the various views encoded in certain subgenres or individual texts. Nonetheless, some broad observations have been put forward. In *The Structure of Scientific Revolutions*, Thomas Kuhn argues that scientific textbooks, which he groups together with popularizations in this context (136–7), display "a persistent tendency to make the history of science look linear or cumulative," erasing all evidence of the paradigm-shifts he sees as integral to scientific advance (139). The reader receives the impression that "science has reached its present state by a series of individual discoveries and inventions that, when gathered together, constitute the modern body of technical knowledge. From the beginning of the scientific enterprise, a textbook presentation implies, scientists have striven for the particular objectives that are embodied in today's paradigms" (140). Steve Fuller makes the same observation more critically: "it is now widely conceded that the most pervasive accounts of the history of science—those found in science textbooks and science popularizations—are misleading to the point of being self-serving ..."; it is left to professional historians of science to "accentuate the blind alleys, dirty deals, and strategic omissions that have enabled the history of science to appear seamless and progressive" (*Science* 80, 19). Thus despite their reintroduction of personal and situational details, popularizations, like scientific papers, tend to reflect an idealized version of the progress of science.

Creative writers have noted the tendency of popularizers to favour smooth, linear narratives, and have used this convention to reflect metafictionally on their own texts. In Janet Turner Hospital's novel *Charades*, Koenig, a physicist working on the Big Bang theory and also an occasional popularizer, resents the convoluted interior narrative of the novel of which he is a part: "'I know it's not logical for a

---

3   There are, of course, exceptions to this general observation (see Lightman, "The Story of Nature").

physicist, of all people, but I have this old-fashioned craving for a simple narrative line. Time curves, it can't be helped, but I don't see why plots should'" (189–90). Unlike the postmodern novelist, the science writer (or at least the stereotypical one) prefers straightforward, apparently naturalistic narratives that do not draw attention to their own narrative status, so that the tale seems more important than its method of telling.

Popularizers themselves are not unaware of this typical linear narrative structure, and occasionally explicitly identify their use of it. In *QED*, Richard Feynman interrupts his potted history of the theories leading up to the development of quantum electrodynamics to give the following qualification:

> By the way, what I have just outlined is what I call a "physicist's history of physics," which is never correct. What I am telling you is a sort of conventionalized myth-story that the physicists tell to their students, and those students tell to their students, and is not necessarily related to the actual historical development, which I do not really know! (6)

Leon Lederman in *The God Particle* similarly alerts his reader to the "'fake history'" told by scientists:

> Scientists (certainly this scientist) use history as part of pedagogy. "See, here is a sequence of scientific events. First there was Galileo, then Newton and this apple ..." Of course, that isn't the way it happens. There are crowds of others who help and hinder ... . However, from the point of view of storytelling, myth-history has the great virtue of filtering out the noise of real life. (412)

Such observations are generalizations, of course, and some popularizers are more willing than others to admit disorderly elements into their narratives of scientific advance. Hawking, who generally does present scientific advance as a smooth progression, at one point in *A Brief History of Time* admits that the 1919 relativity proof was a result of "sheer luck, or a case of knowing the result they wanted to get, not an uncommon occurrence in science" (32). Weinberg observes that the history of research into the early universe "is a rich story for the historian of science, filled with false starts, missed opportunities, theoretical preconceptions, and the play of personalities" (*The First Three Minutes* 18). John Gribbin's *In Search of Schrödinger's Cat* includes descriptions of instances in which new scientific ideas were not immediately incorporated because they did not fit a pre-existing theoretical framework (24, 96, 115), or because the interest of the scientific community was focused elsewhere (17, 42). Gribbin also demonstrates that significant scientific advance may be initially achieved via a makeshift approach bearing little relation to any scientific method: "In 1900 [Planck] made the breakthrough, not through a cool, calm and logical scientific insight, but as an act of desperation mixing luck and insight with a fortunate misunderstanding of one of the mathematical tools he was using" (37–8). In the same book, however, Gribbin admits to smoothing over other awkward details: "To tell a coherent story, I have to make the account more orderly than science itself was at the time ..." (85). Admissions such as these have the advantage of alerting the reader to the constructed nature of the history being delivered, but they fall short of identifying exactly what agenda is informing the constructed narrative aside from a desire for readability. Suggestions that readers simply "love a good story" disguise underlying ideological presumptions; as critics within science studies and science communication are increasingly stressing, it is naïve to dismiss the narrative structure of science popularizations as mere pedagogical scaffolding.

Complementing the broad observations about the linearity of the popular science narrative quoted above are analyses of the different narrative strategies of specific subgenres and individual texts. Greg Myers and Bernard Lightman, for example, have both examined popularizers' use of particular narrative forms (Myers, *Writing Biology* and "Making a Discovery"; Lightman, "The Story of Nature"). Myers emphasizes that by adopting a new narrative form, popularizers produce a new kind of knowledge: "popularizations do not simply transmit or water down the writing of professionals, they transform scientific knowledge as they put it in new textual forms and relate it to other elements of non-scientific culture" ("Science for Women" 171). In his monograph *Writing Biology* he compares popular and professional articles by the same author describing the same research, and concludes that "[t]he professional articles create what I call a *narrative of science*; they follow the argument of the scientist," whereas "[t]he popularizing articles ... present a sequential *narrative of nature* in which the plant or animal, not the scientific activity, is the subject ..." (142). He notes that these specific observations might not transfer directly to popularizations of other scientific fields in which the objects of study are less familiar: "It may be that research on subatomic particles is harder to fit into a narrative of nature than research on butterflies and plants" (188). However, his broader observation—that popularization involves narrative transformation, and constructs "particular views of science" (142)—holds across disciplines.

Physicist Martin Eger, one of the contributors to the multi-disciplinary anthology *The Literature of Science*, has identified a particular type of popular science narrative that spans a number of scientific disciplines. He argues that a range of popularizations, including "Weinberg's much-quoted little book, *The First Three Minutes*," tell "*one and the same story*" (196; 191). This story, which he terms the "new epic" of science, is one of evolution: "evolution explicated in greater detail than ever before, deepened, unified, extended far beyond biology—'universal' or 'cosmic' evolution" (191). More recently Jon Turney has developed Eger's claims in his article "Telling the Facts of Life: Cosmology and the Epic of Evolution." Turney notes that sciences such as cosmology, geology, evolutionary biology and paleoanthropology necessarily create narratives because their "business ... is to reconstruct a timeline" (227). Popularizations of these sciences, he observes, "At their most sweeping ... bring cosmology together with evolutionary theory to construct a linear narrative of the emergence of life in the universe" (228). Like Eger, Turney notes that these "epics" of science address questions which "clearly overlap with those traditionally answered by myth" (230). In another article, focused primarily on Brian Greene's popular physics book *The Elegant Universe*, Turney notes briefly the way in which different narratives can become "intermingled" in popular science books, observing that *A Brief History of Time* "combines a narrative of the universe since the Big Bang with a narrative of the author's own life in science *and* a story of the history of ideas in cosmology that places Hawking in succession to Copernicus, Newton and Einstein" ("Accounting for Explanation" 334).

Building on these previous analyses, and using Turney's last remark as a springboard, I want to examine the relationship between scientific and mythic narrative in *The First Three Minutes* and *A Brief History of Time*. These texts share much in common. Both originated, according to their authors, from presentations given at Harvard University; they deal with similar subject matter, and Weinberg's is the only previous popularization that Hawking mentions (approvingly) in *A Brief History of Time* (vi). Both are explicitly designed as popular texts aimed at non-scientific readers, but neither is an easy read. Hawking crams into one or two chapters theories (such as quantum mechanics or general relativity) that other popularizers spend whole books explaining; Weinberg, although less ambitious in his scope, uses a more technical style, often avoiding the use of equations by simply writing them out as sentences.[4]

These two texts have something else in common: both are concerned to distinguish scientific cosmology from cosmological myths, but both can themselves be read as mythic narratives which see the development of science as the ultimate purpose of the universe. Hawking and Weinberg not only intersperse the story of the evolution of the universe with the story of the evolution of science itself, they also imply that one is the inevitable and fitting result of the other. In *A Brief History of Time*, furthermore, the history of the universe is conflated with Hawking's personal history in a way that reinforces his popular image as the pinnacle of scientific achievement. Before discussing this in detail, I must first outline the myriad arguments about the relationship between science, myth and cosmology that form the context for these popularizations.

Scientific and Mythic Cosmologies: Negotiating the Boundaries

Feynman and Lederman both liken the linear history of physics found in many popularizations to myth: a "myth-story" or "myth-history." But what is it about the narratives to which these physicists refer—the seamless sequential histories told over and over again in popularizations and textbooks—that provokes the word "myth"? These narratives do not have several features that this word might bring to mind: they do not involve Gods or other supernatural figures; they have no explicit religious significance; they are not connected with ritual in any obvious way. Yet clearly, for both popularizers, these modern, secular stories are in some way mythic.

The difficulty of pinning down the exact sense in which Lederman and Feynman apply the term "myth" is a reflection on the term itself. "Myth," like "metaphor," has generated numerous attempts at definition and theorization in a range of different disciplines. A recent attempt to collect the most important writings on myth ran to six volumes, and included contributions from psychology,

---

4 For example: "The number of things of any kind in a fixed volume is inversely proportional to the cube of their average separation, so in black-body radiation the rule is that *the number of photons in a given volume is proportional to the cube of the temperature*" (69).

anthropology, folklore, philosophy, religious studies and literary criticism (Segal). One critic, in an essay entitled "The Problem of Defining Myth," lists ten different overlapping definitions of myth that scholars have employed, such as myth as a "source of cognitive categories," "form of symbolic expression," "projection of the subconscious," "charter of behaviour," "legitimation of social institutions" and "religious communication" (Honko 46–8). Critics working within a multi-disciplinary framework thus usually decline to give a definition of myth, preferring to consider it in terms of a complex of possible characteristics (e.g. Coupe 5–6). Bruce Lincoln begins his monograph *Theorizing Myth* by observing, "It would be nice to begin with a clear and concise definition of 'myth,' but unfortunately that can't be done." Lincoln does, however, venture two "preliminary observations." The first is that myth "regularly denotes a style of narrative discourse and specific instances thereof." The second is that "whenever someone calls something a 'myth,' powerful—and highly consequential—assertions are being made about its relative level of validity and authority vis-à-vis other sorts of discourse" (ix). Both of these observations are central to my analysis of popular cosmology as mythic narrative.

Critics who work within a particular disciplinary framework provide definitions of myth which relate most closely to their own interests. It is the narrative nature of myth which literary critics like myself foreground (Segal viii). Thus when one turns to literary critical sources that are obliged to distil decades of academic debate into pithy definitions, such as the *Concise Oxford Dictionary of Literary Terms*, this is what is emphasized:

> myth, a kind of story or rudimentary narrative sequence, normally traditional and anonymous, through which a given culture ratifies its social customs or accounts for the origins of human and natural phenomena, usually in supernatural or boldly imaginative terms .... In most literary contexts ... myths are regarded as fictional stories containing deeper truths, expressing collective attitudes to fundamental matters of life, death, divinity, and existence .... (Baldick 143).

It is this definition that I will follow here, although I assume that the second "or" in the above definition can also act as an "and": that myths that give accounts of natural phenomena can simultaneously ratify social customs. The space provided by Baldick's qualifiers—"normally" and "usually"—is also important: popular scientific narratives are not themselves "traditional and anonymous" and do not deal with "supernatural" phenomenon. I will argue that the narratives examined below nonetheless draw on elements of traditional, collective mythic narratives and archetypes, thus imparting a mythic quality to science and, in Hawking's case, the scientist.

The particular myths I focus on here are those that can be termed cosmological. According to the *Oxford English Dictionary*, the word "cosmos" suggests not simply the universe, but the universe as a system characterized by order and harmony. Cosmological myth can be considered myth that attempts to give order of some kind to the universe, addressing questions about its beginning, its end, its structure, its purpose and humanity's role in it. One important form of cosmological myth is cosmogonic myth, which deals with beginnings: "all those stories that recount how the world began, how our era started, how the goals that we strive to attain are determined and our most sacred values codified" (Honko 51). Clearly, areas of science such as evolution and cosmology deal with cosmogonic description in its most literal sense, and this has been a source of historical debate between religion and science.[5] However, it is the claim that science also fulfils the further functions of myth—defining society's goals and codifying its sacred values—which lies at the heart of present debate.

Although one can adopt a provisional definition of "myth" for the purpose of a specific analysis, as I have done above, it is important to remember that the term is not always used or understood in this way. It is helpful to distinguish between two broad attitudes towards the subject that have prevailed in the past, and that influence contemporary usage. These can be identified as the "rationalist" and "romantic" versions of myth: "in the first, a myth is a false or unreliable story or belief ... while in the second, 'myth' is a superior intuitive mode of cosmic understanding" (Baldick 143). The rationalist version, which characterizes nineteenth-century approaches, conceives of myth as a form of pre-scientific or proto-scientific knowledge. Sir James Frazer, when he spoke of myth, referred to "mistaken explanations of phenomena, whether of human life or of external nature," explanations resulting from a curiosity which "at a more advanced stage of knowledge seeks satisfaction in philosophy and science" (qtd. in Bascon 26). Thus myth in the rationalist sense is automatically opposed to science; like literature, it is one of the "others" against which science defines itself. By contrast, more recent understandings of myth (such as the one which informs Baldick's definition) have moved closer to the romantic version, focusing on "the structure and function of myth," and emphasizing the need to understand myths within their cultural contexts (Dundes 3). For twentieth-century commentators such as philosopher of science Stephen Toulmin, "it is not enough to regard the old stories only as half-baked science" (23). Arguing by way of example, Toulmin notes that the Atlas myth

> is often thought of as showing only how ignorant the Ancients were: had they known a little more about the solar system, we feel, they would have seen that there was no need for an Atlas ... . Yet surely Atlas was the product not merely of ignorance. There were a vast number of things besides the mechanism of the solar system of which the Ancients were ignorant; but very few of them gave rise to myths. Only where this ignorance was of importance, where it seemed to mean insecurity, was a myth born ... . The stability of the Earth becomes a symbol for so much else. (69)

Myths, Toulmin suggests here, are not solely attempts to explain the physical world, and thus should not be considered redundant or outdated when an apparently better physical model is available. Rather, myth is a literary mode of understanding the world, and the figure of Atlas in this particular myth must be read symbolically rather than literally. Just as metaphor is a literary device that uses an imaginary identity to explore relationships between different objects or ideas, so myth is an imaginary

---

5 Hawking gives one example when he recounts Pope John Paul II's advice that cosmologists should avoid investigating "the big bang itself," adding that he had "no desire to share the fate of Galileo" (116). See Brooke for a historical account of the complex relationship between religion and science.

narrative that allows societies to construct an understanding of themselves and their relationship with the world surrounding them, thereby providing them with (among other things) a sense of security, stability and reassurance. The "myth-histories" identified by Lederman and Feynman can be read in a similar way: as stories designed to give science students a reassuring sense of their place in a stable, smoothly progressing lineage.

While this wider conception of myth removes the sense that it competes on the same terms with science as a description of the world, the science/myth distinction has nonetheless been an unstable and contested one in recent decades. Constructivism's refusal to accept science's epistemological uniqueness allows the argument that science and myth are two equivalent means of understanding and categorizing experience. Damian Broderick observes that, in the constructivist view, "science is finally the kind of story which industrial and post-industrial sophisticates tell about the universe and the creatures which inhabit it, including its story-tellers. Its laws are those not of a special 'scientific method', sought for so long by anxious philosophers, but of narrative and myth" (84). It is unsurprising, then, that myth became one of the terms around which the "Science Wars" pivoted, its looseness of definition exacerbating cross-disciplinary misunderstandings. Alan Sokal and Jean Bricmont are typical of critics of the constructivist view when they complain that the "'postmodern'" philosophy embraced by "[v]ast sectors of the humanities and the social sciences" is characterized by "a cognitive and cultural relativism that regards science as nothing more than a 'narration', a 'myth' or a social construction among many others" (1). This is certainly one context in which Lincoln's second observation—that the use of the word "myth" involves assertions about its validity relative to other discourses—holds true.

Among those quick to defend the unique epistemological status of science against the encroaching claims of myth have been scientist-popularizers. Lewis Wolpert in *The Unnatural Nature of Science* observes "an unwillingness among some anthropologists to regard thinking in primitive societies as somehow inferior to that which characterizes the West," and insists that "[t]he anthropological explanations of cosmologies which reflect the structure of society are very different from the scientists' cosmology, which tries to explain the universe without reference to human beings" (119). Paul Davies, in a 1996 article in the *Times*, criticizes "the mantra of cultural relativism" in which "[s]cientists are presented as little more than myth-makers, whose theories are successful merely on their own terms" ("The Arts Have Lost It" 12). Richard Dawkins writes scathingly in *A River Out of Eden* of "a fashionable salon philosophy called cultural relativism which holds, in its extreme form, that science has no more claim to truth than tribal myth: science is just the mythology favored by our modern Western tribe" (35).

As demonstrated in Chapter 3, however, the position Dawkins and others attack is often more a straw target which can be conveniently dismissed than an accurate summary of the complex and various views held by those working within science studies. Hilary Rose notes that "there is a tremendous range of positions available within the social studies of science" (90), and anti-constructivist "science warriors" tend to respond only to the most easily rebutted. Rose criticizes sociologists of science Harry Collins and Trevor Pinch for their casual use of "the word *myth*," as it is this kind of "ontological slippage" which gives scientists like Dawkins ammunition. She emphasizes that "it does not follow that because scientific claims are socially shaped they are interchangeable with myths or even stories ... . Latour and [the scientists who oppose him] both know that while science is always social, it is not only the social which writes science" (89–90, 95). Bruno Latour, a familiar target of anti-constructivist sallies, has acknowledged the "enormous consequences of science," noting "One cannot equate ... the story telling of origin myths somewhere in the South African bush and the Big Bang theory" (2).

While scientist-popularizers combat myth's perceived encroachment on science's territory, researchers in the humanities and social sciences have pointed to the opposite phenomenon. Toulmin in a 1957 article entitled "Contemporary Scientific Mythology" asserts that "the popular scientist has won over the audience of the popular preacher," arguing that "it is because our contemporary myths are scientific ones that we fail to acknowledge them as being myths at all" (21, 24). He suggests that "Scientific Myths" should be looked for if "theories are regularly invoked in support of conclusions of a kind to which, as scientific theories, they have no relevance; further, if these conclusions are of a sort with which mythologies have from the earliest times been concerned" (27). Pointing particularly to physical cosmology, he notes scientists' tendency to produce "utterances strangely confident and un-tentative" about "the remotest past, and ... the remotest future," and questions the source of this confidence (34–5). Just as, he argues, it is an "anxious fear" about the stability of the Earth that leads to the Atlas myth, it is "[a]n anxious fear of the remote and unknown past, and of the remote and even more unknown future, that leads us to look for eschatological morals even where there is no hope of finding them—in physical cosmology" (70). In particular, he identifies a form of evolutionary myth (specifically in the writings of Julian Huxley). The first step in this myth is to establish humanity as the ultimate product of evolution:

> Then tack on to the beginning of this historical sequence a series of physico-chemical events leading up to the appearance of the first living creatures, and at the latter end treat the development of civilization and technology as a continuation of this biological trend "by other means" ... . Finally, christen your conceptual artefact "the cosmic process" and present it as a golden thread leading from the remotest past up to the present day and on into the future. (61–2)

For Toulmin, this kind of rhetoric suggests a blurring between science as a description of the physical universe and as a mythic system for making sense of the universe.

Much the same argument has been put forward more recently by critics looking at the discourse of popular science. Thomas Lessl, in a 1985 analysis of the rhetoric of Carl Sagan in the television programme *Cosmos*, argues that scientists "promote their status" by turning to "a rhetoric often more characteristic of religious than scientific discourse" ("Science and the Sacred Cosmos" 175). One example, he argues, is Sagan's construction of an "evolutionary myth" in which "[c]osmic evolution, beginning with the 'Big Bang,' begets chemical evolution, which begets biological evolution, which begets human evolution, which begets scientific evolution" (177, 178). In the same way that "Biblical and tribal religions" trace their genealogies back to a sacred time, so in evolutionary myth "[s]cience is the crowning feature of evolution and is endowed with natural authority as the descendent of the cosmos" (178). Fifteen years later, Turney made very similar arguments in his article "Telling the Facts of Life," which I have discussed above.

In the 1990s, at the height of the popular science boom, arguments of this kind became part of the renewed "two cultures" debate in Britain (discussed in Chapter 2), with commentators such as journalist Brian Appleyard and philosopher Mary Midgley criticizing science's perceived attempt to install itself in the position previously occupied by religion. In a 1990 *Sunday Times* article, Appleyard claims that "the new physics in popular mythology *is* the new religion," and that *A Brief History of Time* delivered the message "that science will save us, that physicists are a new priesthood." He concludes, "Science is a myth and Hawking's science is a heroic one" ("God and the Scientists"). In his later book *Understanding the Present* Appleyard expands on this argument, claiming that science continually oversteps its limited domain: it is "incapable of co-existence" and insists on answering questions "*as if* it were a religion" (228). He argues that popularizers such as "Bronowski, Sagan, Hawking, Feynman and Hofstadter" construct a "new spirituality of science" which "arises from our straightforward acceptance of the progressive, evolutionary vision that science provides," and "says that the classical, scientific project is our destiny and we construct our spiritual identities in relation to that destiny" (230). Readers who turn to popularizers like Hawking for metaphysical "succour" find only "the old scientific sleight of hand that makes effectiveness seem like truth" (248). Science accomplishes this sleight of hand by "[beginning] by saying it can only answer *this* kind of question and [ending] by claiming that *these* are the only questions that can be asked"; thus it becomes "the quasi-religious repository of all our faith defined by the popularizers" (249).

Midgley in *Science as Salvation* similarly argues that science is being used to fulfil mythic functions for which it is unsuited. She states that "myth-making" is a "vital human function," but "[t]he way we use science for this function" is not "an acknowledged academic topic" due to a lack of interdisciplinary exchange. Midgley is particularly concerned with the pervasiveness in human thought of teleology— "reasoning from purpose" (9). Notions that the universe must have some kind of goal have been introduced, Midgley argues, "in a rather furtive, unofficial sort of way—in popular books and last chapters" (66). She criticizes a number of popularizers, including both Hawking and Weinberg, but her main targets are the popularizers of the "Cosmological Anthropic Principle"—in particular, Freeman Dyson, Frank Tipler and John Barrow. The Principle, which argues that "the physical universe can in some ways be explained by assuming that it must be such to contain people," becomes in the rhetoric of these popularizers a "value-judgement about the supremacy of knowledge" (27, 67), a claim that "[t]he universe aims to become complete, but cannot do so until it is completely known by people" (26). Their visions of "the indefinitely increasing future glory of the human race" are based on an extrapolation of the theory of evolution, a kind of "hyper-evolution" which extends to "cover the whole development of the universe from the Big Bang onward to the end of time" (147). These authors "form part of a peculiar culture, apparently scientific in its subject matter but highly emotive in its tone, which links the more solemn areas of science-fiction with the more mythical aspects of popular science" (153–4).

These are some of the challenges to the science/myth distinction which scientific cosmology, particularly in its popularized form, faces from critics within the humanities. However, it also faces similar challenges on scientific grounds. Consider the following assertion: "The Big Bang is a myth, a wonderful myth maybe, which deserves a place of honor in the columbarium which already contains the Indian myth of a cyclic Universe, the Chinese cosmic egg, the Biblical myth of creation in six days, the Ptolemaic cosmological myth, and many others" (qtd. in Lerner 228). The source of this claim is Hannes Alfvén, winner of the Nobel Prize for Physics in 1970. Alfvén is no radical relativist. On the contrary, he is a fierce advocate of empiricism, and his claim is specific to Big Bang cosmology. In a paper entitled "Cosmology—Myth or Science?", originally published in the *Journal of Astrophysics and Astronomy* in 1984, Alfvén argues that "the triumph of science" has paradoxically given rise to "a 'scientific creationism' inside academia itself" (594). Alfvén claims that in the face of mounting evidence against the Big Bang theory, cosmologists have not abandoned this theory; instead, he contends, "An increasing number of *ad hoc* assumptions are made, which in a way correspond to the Ptolemaic introduction of more and more epicycles and eccentrics" (596). Thus Alfvén is effectively arguing that Big Bang cosmologists are not adhering to Popperian falsification but rather supporting W. V. Quine's claim that "[a]ny statement can be held true, come what may, if we make drastic enough adjustments elsewhere in the system" (Quine 17). In Alfvén's view, the source of Big Bang cosmologists' willingness to ignore observational evidence is their veneration of what he terms "*mathematical myth*" (593). This myth, which has its source in the philosophy of Plato and Pythagoras, and was resurrected in the twentieth century by Einstein's relativity theory, maintains, "It is possible to explore the structure and evolutionary history of the universe by pure theoretical thinking without very much contact with observations" (599).[6]

Alfvén's criticism of the overly theoretical nature of current cosmology has been echoed by various commentators, such as philosophers of science Paul Feyerabend (137–8) and Mary Hesse ("Cosmology as Myth" 51), and numerous popularizers.[7] Hesse notes the "aesthetic appeal" of science, pointing particularly to "fundamental

---

6 Alfvén's criticisms (like Weinberg's and Hawking's popularizations) predate the discovery in 1992 by the Cosmic Background Explorer Satellite (COBE) of "ripples" in the Cosmic Background Radiation—a discovery which strengthened the Big Bang model. However, John Boslough notes that, despite the COBE discovery, "problems and inconsistencies continued to confront the big bang model" (*Masters of Time* 47). For an interesting perspective on this matter, see Massimiano Bucchi's case study of the reporting of the COBE results in the British media, in which he argues that the publicity given to the results "provided in the first place an opportunity to evade increasing criticism at the specialist level by shifting to a level where Big Bang orthodoxy was still strong and appealing, a level almost inaccessible to the subtleties of critics and alternative models" (98); Bucchi also touches on a number of the broader issues discussed in this chapter.

7 See Lindley 131; Horgan, Chapters 1, 3 and 4; Boslough 54–56; Morris 221; Lerner, Chapters 1–4, 8–10; and Oldershaw, "What's Wrong with the New Physics?" See also

physics and cosmology" (50). This aesthetic appeal, she argues, hints at the function science has taken on: "… whatever other significance scientific theory has, it certainly has the status of cosmological myth in our society, as can be seen in the way 'origins' are taught in schools, and in the popularity of media presentations of fundamental science, both of physics and biology" (51). Some popularizers appear not only to recognize the importance of "aesthetics" in modern cosmology and particle physics but to revel in it. Consider Lederman's *The God Particle*, in which the author holds an imaginary conversation with Democritus:

> DEMOCRITUS: … tell me: you're an experimenter. What physical evidence have you amassed so far for this Higgs particle?
> LEDERMAN: None. Zero. In fact, outside of Pure Reason, the evidence would convince most sensible physicists that the Higgs does not exist.
> DEMOCRITUS: Yet you persist.
> LEDERMAN: The negative evidence is only preliminary. (58)

Similarly, Weinberg in his popularization *Dreams of a Final Theory* explicitly aims to "counteract" what he considers a widespread "over-empiricist" view of science, emphasizing instead "the remarkable power of the physicist's sense of beauty acting in conjunction with and sometimes even in opposition to the weight of experimental evidence" (101, 103). He bluntly states that "we would not accept any theory as final unless it were beautiful" (131).

Creative writers who draw ideas from physics popularization have clearly been struck by this emphasis on aesthetics in cosmology and particle physics. McEwan's protagonist in *Enduring Love* describes how relativity theory was widely accepted and promoted by physicists before its predictions were satisfactorily verified because "it was too beautiful to resist," and how conversely quantum electrodynamics gained acceptance only slowly as it was "unattractive, inelegant. … Acceptance withheld on grounds of ugliness" (49, 50). Turner Hospital's novel *Charades* opens with the cosmologist Koenig writing a popular address: "*The grand unified theories … are difficult to verify experimentally. Nevertheless, they illuminate our understanding of elementary-particle interactions so elegantly that many physicists find them extremely attractive.*" Another character, a non-scientist, considers this sentence "extraordinary": "'Elegance as scientific methodology?'" she queries (5).

Big Bang cosmology, then, along with the fundamental particle physics that is an integral part of it, represents a site at which the relationship between the "two cultures" is being contested by a number of commentators in different fields with different agendas. Constructivists are accused of reducing science to "just another myth"; scientist-popularizers are accused of turning science into a modern mythology; Big Bang cosmologists and particle physicists are accused of valuing symmetry, elegance and beauty over empiricism, producing myth instead of science. It is not my aim here to attempt to reconcile constructivist and realist arguments about science's epistemological uniqueness, or to quell the continuing hostility between the British literary and scientific intelligentsia; nor is it possible for me, as a

---

Oldershaw, "The New Physics—Physical or Mathematical Science?" for a similar criticism of Big Bang cosmology and fundamental physics in a less popular format.

non-scientist, to judge the validity of claims such as Alfvén's. I have discussed these issues primarily to provide a context for the following analysis of Hawking's and Weinberg's popularizations. I do not claim that Hawking's and Weinberg's scientific cosmology, or science in general, is a form of myth. I do claim, however, that in certain contexts, and through certain representations, individual scientific theories and science itself can do the work of myth. Hawking and Weinberg are, on the one hand, interested in explaining scientific cosmology as clearly as possible, and one of the ways in which they do this is by emphasizing science's distance from myth. On the other hand, the narrative structures of their popularizations encode a mythologizing of science.

My argument draws on previous research looking at the concept of "boundary work." Thomas Gieryn explains this concept as "the discursive attribution of selected qualities to scientists, scientific methods, and scientific claims for the purpose of drawing a rhetorical boundary between science and some less authoritative residual non-science" (4). This boundary work takes place, Gieryn explains, in a wide variety of contexts, including "[l]egislative and judicial forums," "the media" and "corporate boardrooms" (24). Felicity Mellor has examined popular physics books as one such site of boundary work. Mellor notes that popular science books claim "merely to set out science in clear and understandable language" while implicitly "identifying what counts as science and simultaneously what does *not* count as science" (512). She analyses popular physics books that highlight the relationship between science and science fiction, thereby working "the boundary between science and fiction," as well as other boundaries, such as those with "religion, magic and technology" (513). In the following, I examine the work that Hawking's and Weinberg's books perform to maintain a boundary between scientific and mythic cosmology, even while they borrow aspects of the latter in the structure of their own narratives.

Beginnings

Both of the popularizations I focus on here are concerned with beginnings. The title of *The First Three Minutes* signals Weinberg's subject: the physics of the very early universe. Hawking's "brief history" promises to cover a greater span of time, but investigations into the Big Bang—the singularity through which the universe came into being—form a central topic of the book. One of Hawking's claims is that the universe may have had no beginning—that it forms a kind of closed, self-contained loop (*A Brief History* 116). This model, as Hawking notes, brings into question the assumption that the universe must have been created by a God-figure (141). Texts, however, have no such option. Books must always have authors (whether they are known or not), and, as much as writers such as James Joyce in *Finnegans Wake* may try to create a closed textual loop, a book must always have a beginning and end. It is at these two points that the text as a narrative "created" or "constructed" by an author is most apparent. The way Hawking and Weinberg begin their narratives of beginning, then, is worth examining in detail.

Popularizer Dennis Overbye, in his *Lonely Hearts of the Cosmos*, identifies the standard convention with which books of his genre open: "It's traditional for a book about cosmology to start by recounting the colorful creation myth of some ancient or primitive society, perhaps partly to show how far we've allegedly come" (2). It is not hard to find examples.[8] Sagan begins his bestseller *Cosmos* with a quotation from an Assyrian creation myth, noting that "[o]ur ancestors were eager to understand the world but had not quite stumbled upon the method," whereas "[t]oday we have discovered a powerful and elegant way to understand the universe, a method called science" (xii). Asimov's *In the Beginning* starts with a juxtaposition of the content of each line of the Book of Genesis with the "scientific view that pertains to the passage" (2). The opening paragraphs of *The First Three Minutes* and *A Brief History of Time* follow suit: both popularizers outline a traditional myth before dismissing it in favour of the scientific world-view. As Turney has noted, the implication in both cases is "one of displacement: of superstition and stories of supernatural forces by a scientific materialist account of the universe" ("Telling the Facts of Life" 230). While in both cases the central problem identified in the myth—infinite regress—is recognized as a problem also facing science, science is nonetheless presented as a superior knowledge system that renders myth redundant.

Weinberg begins *The First Three Minutes* by describing a Norse myth:

> The origin of the universe is explained in the *Younger Edda*, a collection of Norse myths compiled around 1220 by the Icelandic magnate Snorri Sturleson. In the beginning, says the *Edda*, there was nothing at all. "Earth was not found, nor Heaven above, a Yawning-gap there was, but grass nowhere." To the north and the south of nothing lay regions of frost and fire, Niflheim and Muspelheim. The heat from Muspelheim melted some of the frost from Niflheim, and from the liquid drops there grew a giant, Ymer. What did Ymer eat? It seems there was also a cow, Audhumla. And what did *she* eat? Well, there was also some salt. And so on.
>
> I must not offend religious sensibilities, even Viking religious sensibilities, but I think it is fair to say that this is not a very satisfying picture of the origin of the universe. Even leaving aside all objections to hearsay evidence, the story raises as many problems as it answers, and each answer requires a new complication in the initial conditions. (13)

In this passage Weinberg judges the *Edda* solely on scientific criteria. He deliberately misreads the mythic account, treating it as a purely physical, rational description of the origin of the universe, referring facetiously to "hearsay evidence" and "initial conditions." This presentation seems designed to encourage amused contempt on the part of the reader. The cavalier tone of his apology—"I must not offend religious sensibilities, even Viking religious sensibilities"—does nothing to counteract this effect. The way is then clear for Weinberg to introduce his "Modern View" of the universe's origin.

---

8   There are also, of course, counter-examples—popularizers who begin their texts by subverting or qualifying the typical opening gambit of their genre. Overbye is one example; another is John Barrow, who begins *Theories of Everything* with a discussion of myth, but is careful to emphasize that myth should not be perceived as a form of proto-scientific knowledge. The "*raison d'être*" of cosmological myth, he states, was "neither to explain observations nor make predictions" but rather "to embroider a tapestry of meaning within which its authors could represent themselves" (8).

Weinberg's initial dismissal of mythic knowledge is qualified somewhat by later remarks: he admits that his sketch of the standard model of the Big Bang shares with the *Edda* "an embarrassing vagueness about the very beginning" (*First Three Minutes* 17); elsewhere he compares and contrasts the scientifically based cyclic model of the universe with Norse myth without rhetorically disparaging the latter (147); and the *Edda* is the final entry in his "Suggestions for Further Reading," where it is described as "another view of the beginning and end of the universe" (180). In light of his opening paragraphs, however, it is difficult not to read this last remark as ironic and patronizing; the other two references are buried in the text and do not have the positional impact necessary to counter the distinction established in the opening paragraphs.

A Brief History of Time also begins by opposing modern science to a traditional myth, in this case a variant of the Atlas myth. Hawking describes a public lecture on astronomy in which an exchange takes place between "a well-known scientist" and "a little old lady." When the scientist has concluded his description of the orbits of the planets and stars, the woman puts forward her own theory: "'What you have told us is rubbish. The world is really a flat plate supported on the back of a giant tortoise.' The scientist gave a superior smile before replying, 'What is the tortoise standing on?' 'You're very clever, young man, very clever,' said the old lady. 'But it's turtles all the way down!'" (1). Hawking's treatment of cosmogonic myth is not so dismissive as Weinberg's. His scientist with his "superior smile" seems poised for a fall; and indeed, while Hawking suggests that most people would find the little old lady's "infinite tower of tortoises" model of the universe "rather ridiculous," he goes on to ask "why do we think we know better?" He follows with a series of questions about the universe—where did it come from, did it have a beginning, and so on—before stating, "Recent breakthroughs in physics ... suggest answers to some of these longstanding questions" (1). He then points out that in the future our current physical models of the universe may seem "as ridiculous as a tower of tortoises" (2). He thus represents science as the answer to the questions implicit in creation myths, while simultaneously highlighting the transitory nature of scientific knowledge.

Although Hawking's approach is more complex than Weinberg's, their opening paragraphs share the same basic function: to allow the reader a chuckle at the expense of mythic thought and thereby to establish the rationality of myth's implicit alternative, science. The humour content of Hawking's anecdote arises more from the inappropriate context in which the "infinite tower of tortoises" is suggested than the myth itself. It is a rhetorical cheap shot: with the mythological world-view reduced to the clichéd (and strategically gendered) figure of "a little old lady" pontificating about turtles in an embarrassingly inappropriate forum, it is not difficult for the scientific view to gain authority in comparison.

The salient feature of Hawking's and Weinberg's representations of myth and science is their desire to judge these two knowledge-representation systems by the same criteria, and hence their willingness to ignore the encoded structures of mythic thought, structures which refer to aspects of human society apart from knowledge of the physical world. They are adhering to the nineteenth-century conception of myth as solely proto-scientific knowledge. The obvious interpretation of the role of the mythic broad-backed tortoise is that it provides its creators with a sense of security: "The stability of the Earth becomes a symbol for so much else" (Toulmin 69). It is this symbolic dimension of myth which Hawking's and Weinberg's opening paragraphs are designed to ignore. Of course, it is not Hawking's or Weinberg's role to embark on an elaboration of the function of myth. But in both cases there is no suggestion that these two forms of knowledge representation should be seen as complementary, as embodying different kinds of information, both of which are useful to humanity; that, as Jean-François Lyotard states in *The Postmodern Condition*, it is "impossible to judge the existence or validity of narrative knowledge on the basis of scientific knowledge and vice versa: the relevant criteria are different" (39).

To see that Hawking and Weinberg could have argued otherwise, it is worth turning for a moment to another popularizer's treatment of the same idea. In a popular address given in 1963 and published in 1998 as *The Meaning of It All*, Feynman attempts to convey to his audience the excitement of science:

> ... the ancients believed that the earth was the back of an elephant that stood on a tortoise that swam in a bottomless sea. Of course, what held up the sea was another question. They did not know the answer.
>
> The belief of the ancients was the result of imagination. It was a poetic and beautiful idea. Look at the way we see it today. Is that a dull idea? The world is a spinning ball, and people are held on it on all sides, some of them upside down. And we turn like a spit in front of a great fire. We whirl around the sun. That is more romantic, more exciting ... .
>
> This universe has been described by many, but it just goes on, with its edge as unknown as the bottom of the bottomless sea of the other idea—just as mysterious, just as awe-inspiring, and just as incomplete as the poetic pictures that came before.
>
> But see that the imagination of nature is far, far greater than the imagination of man. (9–10)

Feynman is also establishing a hierarchy between science and myth, but in basing this hierarchy on aesthetic criteria he is, in a sense, taking myth on its own terms, judging science by the yardstick of myth, rather than the other way around. Although the scientific view is portrayed as more imaginative than the mythic view, and there is an assumption that science is constrained by nature in a way that myth is not, there is not the same suggestion that science is a superior system of knowledge by virtue of this constraint. This reading is supported by Feynman's explicit statements later in the same chapter: "if a thing is not scientific, if it cannot be subjected to the test of observation, this does not mean that it is dead, or wrong, or stupid. We are not trying to argue that science is somehow good and other things are somehow not good" (16); for Feynman, the things "left out" of the scientific method are "in many ways the most important" (17).

Why this need, then, in the cases of Hawking and Weinberg, to devalue myth by portraying it as a kind of poor man's—or, rather, old lady's—science? A possible answer lies in the view that the kind of physics Hawking and Weinberg are promoting itself partakes of the characteristics of myth—that they are at some level eager to emphasize the science/myth division because they recognize its instability in the area of their own research. Certainly, there are suggestions in *The First Three Minutes* and *A Brief History of Time* that the authors are slightly anxious that the science they discuss be recognized as empirically based. Towards the end of his first chapter Weinberg states, "It is a tribute to the essential objectivity of modern astrophysics that this consensus [the 'standard model' of the Big Bang] has been brought about, not by shifts in philosophical preferences or by the influence of astrophysical mandarins, but by the pressure of empirical data" (17–18). Given that in the popular view this is the way in which all scientific research is supposed to proceed, it seems slightly odd that Weinberg needs to emphasize the experimental basis of Big Bang cosmology here. Later, he admits that he might be guilty of "a note of scientific over-confidence," but explains that he does not believe that "scientific progress is always best advanced by keeping an altogether open mind. It is often necessary to forget one's doubts and to follow the consequences of one's assumptions wherever they lead—the great thing is not to be free of theoretical prejudices, but to have the right theoretical prejudices." Weinberg does go on to discuss the importance of experimental evidence, stating that the "test" of such a prejudice is "where it leads," and asserting that the standard model of the Big Bang does provide a "theoretical framework for future experimental programmes" (118). However, his effective admission of the theory-ladenness of science seems to prefigure his criticism of the "overly empiricist" view of science and advocacy of "beauty acting ... even in opposition to the weight of experimental evidence" in his later popularization, *Dreams of a Final Theory* (101, 103). The ambiguous attitude towards the role of empiricism in cosmological research expressed in this passage makes his earlier dismissal of "philosophical preferences" appear revealingly defensive.

Hawking goes to greater pains to emphasize the sound scientific basis of cosmological research. Early in *A Brief History of Time* he declares, "As philosopher of science Karl Popper has emphasized, a good theory is characterized by the fact that it makes a number of predictions that could in principle be disproved or falsified by observation" (10). In his concluding chapter he repeats this point, this time contrasting the "tower of tortoises" myth with superstring theory and deeming the former an unsuccessful theory:

> Both theories lack observational evidence: no one has ever seen a giant tortoise with the earth on its back, but then, no one has seen a superstring either. However, the tortoise theory fails to be a good scientific theory because it predicts that people should be able to fall off the edge of the world. This has not been found to agree with experience, unless that turns out to be the explanation for the people who are supposed to have disappeared in the Bermuda Triangle! (171)

As well as flippantly connecting an ancient creation myth with a more recent (and much ridiculed) legend in order to dismiss the former, Hawking implies that myths are failed scientific theories because their predictions are falsified. This is ironic considering that superstring theory is the one area of physics to which the application of Popperian criteria is *most* problematic. By admitting the trivial point that superstrings are not directly observable, Hawking avoids the more serious problem that superstring theory has failed to make predictions. Nobel Laureate physicist Sheldon Glashow and his colleague Paul Ginsparg, in an article on superstrings in *Physics Today* published two years prior to *A Brief History of Time*, note that "years of intense effort by dozens of the best and brightest have yielded not one verifiable prediction, nor should any soon be expected." They envisage superstring theory evolving into an activity "to be conducted at schools of divinity by future equivalents of medieval theologians," and suggest that "our noble search" may end "with faith replacing science once again" (Ginsparg and Glashow 7).[9] And (as previously noted) Hawking's main subject, the Big Bang theory, does not exactly fit Popper's requirements either. Alfvén and many others contend that Big Bang cosmologists are one group of scientists who do *not* accept falsifying evidence but merely add "an increasing number of *ad hoc* assumptions" to their theory.[10] Within this wider context, Hawking's comparison of superstring theory with the tower of tortoises can be read as an instance of rhetorical boundary work. Hawking does not need to add anything further about the predictions of superstring theory, as he has already shored up its scientific status by opposing it to a clearly non-scientific "other."

It is possible to find other instances where Hawking rhetorically skirts the question of empirical evidence. Near the end of the third chapter, he describes a paper written by Roger Penrose and himself that, he claims, provides a proof of the existence of a Big Bang, given that "general relativity is correct and the universe contains as much matter as we observe." He goes on:

> There was a lot of opposition to our work, partly from the Russians because of their Marxist belief in scientific determinism, and partly from people who felt that the whole idea of singularities was repugnant and spoiled the beauty of Einstein's theory. However, one cannot really argue with a mathematical theorem. So in the end our work became generally accepted and nowadays nearly everyone assumes that the universe started with a big bang singularity. (50)

To begin with, we might note that Hawking claims that *opposition* to the Big Bang was based on extra-scientific principles, including aesthetics—this is Alfvén's argument in reverse. More revealing is Hawking's rhetorical transformation of his proof of the Big Bang, which he initially states is based on both general relativity *and* the observation of matter, into a "mathematical theorem." It may be true that one cannot argue with a mathematical theorem, but one *can* argue about whether

---

9 Feynman too has explicitly criticized superstring theory's non-empirical basis, complaining that researchers in this field avoid experimental disagreement by adding ad hoc hypotheses: "I don't like that they're not calculating anything. I don't like that they don't check their ideas. I don't like that for anything that disagrees with an experiment, they cook up an explanation—a fix-up to say, 'Well, it still might be true'" (qtd. in Davies and Brown 194).

10 Conversely, it can be argued that Alfvén's alternative cosmological research programme, based on plasma physics, has produced *correct* predictions that have had little effect on the scientific community. Stephen Brush, a historian of science and former physicist, has used Alfvén's research as a test case to see whether scientists actually follow the policy of judging theories on the strength of their predictions. Brush concludes that while "[b]y Popperian criteria [Alfvén's] theories should have acquired credit by their successful predictions," this has not occurred; he suggests that "prediction as a means of evaluating scientific theories has been exaggerated" ("Alfvén's Programme" 584, 585).

it is an acceptable description of the universe: Hawking and Penrose's theory also rests on observation, and the amount of matter in the universe is one of the points which Big Bang critics dispute.

Hawking's and Weinberg's textual beginnings can be interpreted, then, as defensive responses to perceived shortcomings in the scientific nature of their own theories. They need to be read in the context of accusations of a lack of empiricism and an overly aesthetic approach that have been made against Big Bang theory, and against later episodes in both texts where the popularizers implicitly defend themselves against such accusations. By opposing scientific and mythic cosmology at the outset, Hawking and Weinberg pre-empt any suspicion on the part of the reader that their own science contains elements of myth. Boundary work of this kind, as Mellor has noted, is not uncommon in popular physics books. What is particularly interesting in *The First Three Minutes* and *A Brief History of Time* is that both narratives themselves have mythic qualities that undermine the myth/science distinction with which they begin. In order to see this, it is necessary to look at how each narrative develops, and how it concludes.

Middles

The "middle" of a narrative is its main body; as both Hawking's and Weinberg's books are dense reads, it is not possible to analyse and compare these sections with the same kind of attention that can be paid to their beginnings and endings. It is possible, nonetheless, to make some broad observations about their structures.

Weinberg's book claims (if the blurb on the back of my Fontana edition is to be trusted) to give a "blow-by-blow" account of the "first three minutes" of the universe.[11] In his introduction, he duly gives a three-page summary of this period, starting with the sentence, "In the beginning there was an explosion" (14). He then turns, in the middle of his narrative, to giving a more detailed account of this history. Unsurprisingly, however, Weinberg is not able to begin this detailed account until he has provided another history: the history of the development of cosmology and particle physics. It is more than one hundred pages into his book before he can announce: "We are now prepared to follow the course of cosmic evolution through its first three minutes" (102). He gives this history not continuously, but in a series of discrete moments, which he likens to film frames, starting at one one-hundredth of a second after the Big Bang. The beginning of the next chapter, however, has him asking his reader to "turn away for a moment from the history of the early universe, and take up the history of the last three decades of cosmological research" (120). He then returns in his last chapter to a narration of the very beginning of the universe—the first one-hundredth of a second—which he follows with a short epilogue that looks towards the future, rather than the past. Broadly, the narrative oscillates between a history of the universe and a history of

---

11 Actually, Weinberg admits, he covers three and three-quarter minutes; he apologizes to his reader for his "inaccuracy," explaining that the simpler title "sounded better" (110). One wonders why he did not more accurately round up to four minutes; perhaps because in Western society "three" has a more mythical resonance than "four"?

research in cosmology, with small sections focused on the former topic sandwiched between larger sections on the latter.

Where Weinberg's title suggests an account of a period of time, Hawking's title offers a history of time itself. What Hawking actually provides is a series of parallel histories of ideas in physics—ideas about space and time, the expansion of the universe, the nature of matter, black holes—followed by a discussion of his own more recent ideas. Unlike Weinberg, who highlights occasions when the history of science did not progress as expected, Hawking tends to rely on what Lederman terms the "myth-history" format: that is, the description of discoveries in physics as a linear series of events leading up to contemporary theories. In chapter 2, for example, the understanding of space and time is seen to progress fairly smoothly through Galileo, Newton, Maxwell and Einstein, to Hawking and his colleague Roger Penrose; in Chapter 5, an initial history of atomic physics moves through Dalton, Thomson, Rutherford, Chadwick and Murray Gell-Mann (63–5). These histories then allow Hawking, in his later chapters, to expound more recent research in which he has been directly involved. In the eighth chapter, he gives a narrative of the "generally accepted" (116) history of the universe similar to that provided by Weinberg, before going on to propose his speculative "no boundary" universe in which there is no beginning or end. He returns to his narrative of science in his tenth chapter, "The Unification of Physics," which is followed by a short conclusion. As in Weinberg's text, one narrative (that of the development of the universe) is structurally contained within another (the development of physics).

This two-part structure is not surprising; it is not possible to describe the early universe, with its soup of fundamental particles, without explaining (among other things) what these particles are, why they are thought to exist, and how they behave. As Myers foresaw, the narrative of the cosmos cannot be told in the form of a straightforward "narrative of nature," the way that a biologist might tell a narrative of a particular animal or plant, as the reader has no direct experience of the "nature" being described. Instead, the narrative of nature must be produced by the narrative of science. What is significant in both texts, however, is that their conclusions reverse this structure, implying that the narrative of nature produces (in a different sense) the narrative of science. This is achieved through the adoption of the "teleological myth" that Midgley and others have identified, in which science becomes the purpose towards which the universe has always been unfolding.

Ends

When literary critics discuss endings, they often divide them into the categories of "closed" and "open" endings. The former, typical of realist novels, tie up the loose threads of the narrative, and give a reassuring sense of completeness. In the latter, more often found in modernist and postmodernist novels, the fate of the characters and the consequences of various narrative events are left hanging; resolution is refused. The same categorization might be applied to models of the end of the universe. A universe that ends with a "bang" (to use T. S. Eliot's words), or a "big crunch" (to use Hawking's), at least has the advantage of a sense of closure. The universe that ends with a "whimper"—endless expansion and cooling—is disturbingly open, continuing indefinitely. Unsurprisingly, the terms "closed universe" and "open universe" are used to describe these two possibilities (Weinberg 115).

Literary critic Eric White, in an article entitled "Contemporary Cosmology and Narrative Theory," compares various scientific views of the universe with the "four primary modes of emplotment": comedy, tragedy, romance and farce. He notes that comedy and tragedy "are both modes of emplotment in which closure is a principal aesthetic objective"; the former mode achieves this via "*resolution* of conflict," the latter via "*revelation* of the limiting parameters of existence" (102). Romance and farce, by contrast, "depict reality as an endless sequence of events" (102). White's main objective is to reconcile the postmodern distrust of metanarratives with the continuing need to give "comprehensible narrative form" to "what would otherwise be a demoralizing incomprehensible flux" (101). He concludes that farce is the "most appropriate way to tell the story of nature" (107): "the farcical vision conceives of history as a process without a *telos* or goal in which promise and possibility oscillate interminably with the prospect of devolution" (108). His main support for this, however, comes not from cosmologists such as Hawking and Weinberg, but from popularizers of chaos and complexity theories. Weinberg, he notes briefly, favours a now-outdated "tragic emplotment" (104).

It is not difficult to see how White draws this conclusion, as Weinberg explicitly employs the categories of narrative knowledge in the concluding paragraphs of *The First Three Minutes*. In his penultimate paragraph, Weinberg notes humanity's urge to believe, in his words, "that human life is not just a more-or-less farcical outcome of a chain of accidents reaching back to the first three minutes, but that we were somehow built in from the beginning" (148). He implicitly rejects this reasoning, presenting a profoundly pessimistic view of the universe as overwhelmingly vast and inhospitable, and facing "a future extinction of endless cold or intolerable heat." He ends the paragraph with his much quoted observation, "The more the universe seems comprehensible, the more it also seems pointless" (148–9). Yet, although this is often presented by commentators as the concluding sentence of *The First Three Minutes*, Weinberg does not end his overall narrative with such nihilistic portents. Rather, in his final paragraph, he succumbs to the teleological urge he has just rejected, and turns to the practice (rather than the findings) of science as a fleeting but nonetheless worthwhile source of purpose:

> But if there is no solace in the fruits of our research, there is at least some consolation in the research itself. Men and women are not content to comfort themselves with tales of gods and giants, or to confine their thoughts to the daily affairs of life; they also build telescopes and satellites and accelerators, and sit at their desks for endless hours working out the meaning of the data they gather. The effort to understand the universe is one of the very few things that lifts human life a little above the level of farce, and gives it some of the grace of tragedy. (149)

Weinberg here returns to the myth/science hierarchy with which he began his text. People are not content with myth—"tales of gods and giants"; they need its implicit successor, science. This propels the story of the universe above the category of farce, which is a "chain of accidents" with no direction, to that of tragedy, in which purpose is embodied in, and "revelation" (to use White's terms) achieved by, the doomed efforts of the tragic hero, humanity. Unlike mythic knowledge, scientific knowledge provides purpose, a *telos*. Weinberg's narrative itself finds closure by suggesting that scientific progress gives closure to the universe, or at least humanity's part in it.

The end that Weinberg gives his narrative needs to be read in the context of its overall structure. In the body of his text Weinberg frames one narrative, the evolution of the universe, with another, the evolution of science. In his conclusion, he implies that the latter narrative is what gives the former its purpose, what elevates it from farce to tragedy. There is an imbalance here, a form of anthropocentrism in which the narrative of cosmological research is seen as equally momentous, on a universal scale, as the development of the universe itself.[12] A text which sets out to give a history of the early universe ends by positing cosmological research as the only endeavour that gives the universe meaning. This is a mythic, rather than a scientific, narrative: it is a story which explains Western scientific society to itself, which justifies its values and practices, and assuages its insecurities.

Hawking also favours a closed mode of emplotment, but it is comedy, rather than tragedy, which he prefers. *A Brief History of Time*, like many popular physics books, turns in its final chapter to the question of "The Unification of Physics": the coming together of various "partial theories" into a "complete unified theory," a theory which would unify the four forces of nature (gravity, electromagnetism, and the strong and weak nuclear forces) (155–6). This is the "Theory of Everything," a common theme in physics popularization as the second millennium drew to a close. Such a theory is, as Weinberg states in his *Dreams of a Final Theory*, a "starting point, to which all explanations may be traced" (3), but it is also popularly perceived as the endpoint of physics. Hawking himself had predicted in 1980 that a TOE would be achieved by the end of the twentieth century; in *A Brief History of Time* he is more cautious, but still believes there is a "good chance" that a TOE will be reached "within the lifetime of some of us who are around today" (167). It is easy to see how this emphasis on final unity corresponds to the comic mode, which, as White states, achieves closure through the resolution of conflict (102)—traditionally, through the union of marriage.

This turn towards unification is something that Hawking foreshadows early in his popularization. Near the end of his first chapter, he asks how we can know that science will come to the correct ultimate conclusion—the correct "complete unified theory" (12). In answer, he draws on "Darwin's principle of natural selection," arguing that "provided the universe has evolved in a regular way, the reasoning abilities that natural selection has given us would be valid also in our search for

---

12 The same imbalance can be found in the title of another Big Bang book, Marcus Chown's *Afterglow of Creation: From the Fireball to the Discovery of Cosmic Ripples*. Here there is an implied equivalence of significance between the bookends of the narrative: Chown's title promises to give a history not of the time between two scientific events—from the *proposal* of the Big Bang *theory* to the discovery of the ripples—but between a natural and a scientific event—from the explosion itself to the COBE results.

a complete unified theory, and so would not lead us to the wrong conclusions" (12–13). He closes the chapter with a vision similar to Weinberg's:

> ... ever since the dawn of civilization, people have not been content to see events as unconnected and inexplicable. They have craved an understanding of the underlying order in the world. Today we still yearn to know why we are here and where we came from. Humanity's deepest desire for knowledge is justification enough for our continuing quest. And our goal is nothing less than a complete description of the universe we live in. (13)

This is teleology with a vengeance: human existence must have a goal, and that goal is knowledge of the universe—science. According to Hawking, science is quite literally the product of the universe's evolution, so essentially this is a vision of the universe achieving its purpose through self-knowledge. It is this kind of teleological myth that Midgley and the other like-minded critics discussed above have recognized as widespread in the discourse of popularizers.[13]

Hawking returns to this idea in the short and now controversial "Conclusion" which follows his chapter on the unification of physics. Here, he leaves little doubt that the quest for knowledge that is humanity's goal is a specifically scientific one. In his penultimate paragraph he implicitly dismisses the role of the humanities in a search for a complete theory, noting with disdain that philosophy, unable to absorb modern science, has been reduced (in Wittgenstein's words) to "'the analysis of language'" (175). A "complete theory," for Hawking, is completely scientific. Although it will potentially enable everyone, "philosophers, scientists, and just ordinary people" to debate the reason for the universe's existence, this can happen only when physicists have finished their search. The discovery of such a theory, he proclaims in his much-quoted last line, would be "the ultimate triumph of human reason," and would allow us to "know the mind of God" (175). In *Mythologies*, Roland Barthes observes that "the mythology of Einstein" and his famous equation $E=mc^2$ allowed the world to regain "the image of knowledge reduced to a formula ... science entirely contained in a few letters ... the ideal that total knowledge can only be discovered all at once" (69). Hawking's conclusion promotes the same ideal: the search for a core centre from which all meaning can be derived—what Jacques Derrida termed a transcendental signified, a *logos*. The "mind of God" remark then identifies the philosophical and theological senses of *logos*. The search for this *logos* is an integral part of research into the Big Bang, and together they provide the *telos* of human existence. Given that literary criticism in the last few decades has been characterized by a suspicion of transcendental signifieds, it is hardly suprising that "analysis of language" is peripheral to this search.

Although Hawking favours the comic narrative mode and Weinberg the tragic, they both achieve closure by presenting science (cosmology and particle physics, in particular) as the activity that gives meaning and purpose to human life. Scientific

---

13 Of course, it is possible to argue that there is a basis for considering scientific success the inevitable product of evolution, but Hawking's paragraph-long treatment of this controversial subject hardly establishes this basis, especially given that some evolutionary biologists (such as Gould) fiercely argue the opposite case.

cosmology as popularized by Hawking and Weinberg, then, fulfils the symbolic functions of myth—it recounts, in Honko's phrase, "how the goals that we strive to attain are determined and our most sacred values codified." Hayden White argues that narrativity imposes upon real events "a coherence that permits us to see 'the end' in every beginning" (23). The narratives of *The First Three Minutes* and *A Brief History of Time* impose a narrative structure upon the universe, so that its beginning—the Big Bang—permits through evolutionary progress its final goal—complete knowledge of itself through science, a Theory of Everything. Weinberg and Hawking tell tales, not of giants and gods, but of the cosmic lineage of the scientific project. We can see Hawking's and Weinberg's teleological narratives as science's attempt to graft its own history onto the evolution of the cosmos, its own purpose onto the purpose of the universe: to mythologize its own origins and ends.

Mythologizing the Physicist

If *A Brief History of Time* and *The First Three Minutes* can be read as mythic narratives that present science as the ultimate purpose of the universe, where does the scientist (and the scientist-popularizer) sit in this discourse? This depends both on the extent to which each scientist inserts himself into the text, and on the extent to which each has developed a profile outside of the text that informs its reading. The next chapter deals with stereotypes of the scientist in popularizations, but it is worth exploring briefly here the way in which the scientist-author can be mythologized along with a narrative.

Steven Weinberg, a Nobel laureate, is well known within the physics community and, thanks to his vocal contributions to the "Science Wars," also within the community of science studies researchers. His is not, however, a household name in the sense that Hawking's is; it is unlikely that the average reader of *The First Three Minutes* would have a mental image of Weinberg or a sense of his public persona. While Weinberg's colleagues and various "science warriors" might relate the mythic narrative of *The First Three Minutes* to Weinberg himself, for most readers he is simply a scientific authority. The text itself does relatively little to encourage any sense of Weinberg beyond this. As a narrator, he makes comparatively few intrusions into his text. He occasionally uses the first-person singular voice, but for the most part remains outside of the narrative. There are only a few places where he refers directly to his own life and research. The first is the preface, in which he explains his own research (mainly particle physics, and "small bits" of cosmology [9]) and the origin of the book (a talk at Harvard), outlines his aims, and thanks his "colleagues in physics and astronomy" and several others (12). He thus establishes early on his authority as a scientist and his purpose as a popularizer. Near the start of the first chapter, and several times in the later chapters, he makes brief references to his own place within the research described (13–14, 98, 135, 138, 139); and occasionally he ventures an opinion, or mentions a communication with another scientist. The most personal intrusion, however, occurs in the penultimate paragraph, when he describes himself, on an aeroplane, writing the paragraph and looking down on the Earth. The reader's last glimpse of the scientist-author thus bestows upon him a God-like perspective. This dovetails with the mythic purpose attributed to science in the final paragraph: given that Weinberg is himself doing the kind of research that he believes the universe has been evolving towards, this positions him (and those like him) as the actor who provides the universe with meaning. But neither this passage, nor *The First Three Minutes* itself, is enough to give Weinberg himself, as a scientist, mythic status within the popular imagination.

Hawking's case is rather different. Unlike Weinberg, Hawking—both his name and image—is instantly recognizable to many readers. His fame preceded the publication of *A Brief History of Time*; as outlined in Chapter 1, Hawking first entered the public consciousness in the late 1970s, when he featured in Nigel Calder's television series and book, *The Key to the Universe*. By the early 1980s, according to his biographers, he was being profiled regularly in newspapers and magazines such as *Newsweek* and *Vanity Fair* (White and Gribbin 205). Many commentators suggest that the spectacular success of *A Brief History of Time* was due as much to Hawking's own image as it was to his text. Certainly the marketing of the book suggests this: popularizers themselves almost never feature on the dustcovers of their books, but Hawking, pictured against a blackboard of equations or the starry cosmos, dominates most editions of *A Brief History of Time*. More than any other contemporary scientist, Hawking is a celebrity, a part of popular culture.

The "Hawking phenomenon" is one aspect of physics popularization that has drawn significant attention from researchers, several of whom argue that Hawking has taken on a mythic role within Western culture.[14] Margaret Wertheim (among others) observes that Hawking's "mystical aura" is "compounded by the extreme disjunction between the power of his mind and the lameness of his body. Here he embodies an archetype found in many cultures around the world—the lame or crippled seer" (218). The fact that the seers of mythology and folklore are often blind or "crippled" can be linked to the belief that removal or marginalization from ordinary, everyday sensory existence allows for closer contact with extraordinary, supernatural forces. The same equation is applied to Hawking. "Even as he sits helpless in his wheelchair," reads the quotation from *Time* magazine on the back of my Bantam Press edition of *A Brief History of Time*, "his mind seems to soar ever more brilliantly across the vastness of space and time to unlock the secrets of the universe."

Yet, just as a mythic narrative is used to dismiss mythic knowledge as science's poor cousin in *A Brief History of Time*, so Hawking's mythologization paradoxically enshrines science as the ultimate source of insight into the meaning of the universe. He is always seen attached to mechanical objects—motorized wheelchair, computerized voice synthesizer—and the attributes of these machines are bestowed upon his intellect. This "cyborg" nature reinforces, rather than contradicts, the sense that he has something akin to mystical insight into the universe. As Barthes observes, myth "could not care less about contradictions as long as it establishes a

---

14 For analyses of Hawking's image, *A Brief History of Time*, and Morris's documentary film of the same name, see Rodgers; Dunning-Davies; Higley; Bushnell; Michael; Rosenheim; Wertheim; and both listed articles by Mialet.

euphoric security." Just as Einstein, according to Barthes, is "at once magician and machine" (70), so Hawking signifies both a mystical connection with the universe and a computer-like rationality. Although Hawking has explained how his disability forced him to develop new pictorial ways of thinking (*Black Holes* 31), for an audience watching him, he has no discernible "method." A question is asked, a number of minutes ensue during which processes mysterious to the audience take place, then the computerized voice proclaims: "I will answer ... ."[15] Hawking has taken on the role of modern oracle: what other speaker could expect an audience to wait so long in such anticipation for his declaration? The equipment to which he is attached, and his popular reputation as the foremost physicist of his generation, suggest to the public that the processes which produce his answers, while inaccessible and opaque, are nonetheless as purely rational and logical as it is possible to be. The myth of Hawking reassures post-Enlightenment Western society that there is a least one person whose thought processes are not (apparently) adulterated by irrational bodily impulses, whose mind roams "free" of these constraints. The fact that Hawking is himself avowedly anti-mystical reinforces rather than contradicts his role as a seer;[16] for what is being mythologized in his image is not religion, but science.

To what extent then does Hawking himself contribute to his own mythologization in *A Brief History of Time*? He has stated that the book "was intended as a history of the universe, not of me" (*Black Holes* 33); but Hawking intrudes into his narrative more frequently than Weinberg. Although, like Weinberg, he uses the first-person singular voice sparingly, he refers to his own life and research in more detail, and these references increase as the text continues. This pattern is followed in miniature in the individual chapters, four of which have concluding paragraphs which make some reference to Hawking's own work. The effect is to place Hawking's research as the end-product of the "myth-histories" of physics that he provides. This effect is exacerbated by three appendices to the text. Although Hawking concludes the narrative proper with his notorious "mind of God" remark, he attaches to the end of his text brief biographies of Albert Einstein, Galileo Galilei, and Isaac Newton. There is no clear reason for including these biographies, which promote these three men above the community of scientists discussed in the book, and imply (as several commentators have remarked[17]) that Hawking himself should be the next name on this list. Comments in the text encourage such a presumption: Hawking tells his reader of his "strong sense of identity" with Galileo, as he was born "exactly 300 years" after Galileo's death (116), and notes that he holds "the same professorship that Newton had once held" (68). Thus Hawking's own narrative as a researcher is grafted onto the mythic narrative identified above in which science (and more specifically the search for a Theory of Everything) is the goal towards which the

---

15 I base this description on Hawking's lecture "The Theory of Everything" in Oxford Town Hall, which I attended on 24 February 1998.

16 For an example of Hawking's anti-mystical stance see his interview with Renée Weber, included in her *Dialogues with Scientists and Sages*.

17 See, for example, White and Gribbin (250); Polkinghorne (*Beyond Science* 50).

universe has been evolving.[18] He effectively places himself as the pinnacle of this cosmic evolution, tracing his genealogy back to a sacred time, as Lessl might argue.

This self-construction was later repeated and reinforced in a different popular context, when Hawking at his own request featured in the science fiction series *Star Trek: The Next Generation* (Okuda, Okuda and Mirek 123). Hawking made a cameo appearance as a computer-generated holographic image of himself, playing poker with similar holograms of Albert Einstein and Isaac Newton; all have been generated by the android Data as an experiment ("Descent").[19] A later episode presents an alternative future in which Data has become the holder of the Lucasian Chair at Cambridge, thus confirming the lineage hinted at in *A Brief History of Time* ("All Good Things"), and taking the image of the physicist as a kind of "thinking machine" to its extreme. In a sense, it is no surprise that Hawking appeared in a science fiction series as a simulacrum of himself, for "Hawking" as the public know him has become a kind of fictional character, a symbol as much as a reality, a mythic figure as much as a man.

What are the consequences of such mythologizing? One commentator, physicist Jeremy Dunning-Davies, argues that "the status Hawking has achieved in popular culture has influenced his status within the scientific culture" (86). Papers that challenge Hawking, he suggests, are rejected "because his reputation has in some sense gone beyond the purely scientific" (86). This is a clear example of the feedback between popular and specialist scientific writing that I mentioned in my introduction. And the same phenomenon is possible, one can speculate, on the level of scientific fields rather than individual researchers. Scientists who are able to mythologize their research through popularizations potentially increase public support for it. Lyotard suggests that although scientists dismiss narrative knowledge as untestable, and consider it "savage, primitive, underdeveloped, backward," when it comes to describing their discoveries to the public, "they play by the rules of the narrative game .... The state spends large amounts of money to enable science to pass itself off as an epic" (27–8). Conversely, passing science off as epic helps to justify the state spending a large amount of money on it. Weinberg complains at one point in *The First Three Minutes* that "truly cosmic investigations" such as those he describes "deserve a large share of the space budget" (75); his later popularization *Dreams of a Final Theory* ends with him roaming the building site of the Superconducting Super Collider, hoping (in vain, as it transpired) that threats to its funding will not eventuate (219–20). Big science requires grand narratives. We can see the closed teleological narratives of popularized Big Bang cosmology and the search for a

---

18 Errol Morris's documentary of Hawking's life and work, also entitled *A Brief History of Time*, quite blatantly parallels Hawking's own history with the evolution of the cosmos. For example, Hawking's discussion of the past and future of the universe is juxtaposed with images from his childhood and interviews with family and friends about his early development. See Bushnell for a discussion of the way both versions of *A Brief History of Time* "conflate cosmological time and personal time" (684).

19 For analyses of Hawking's appearance on *Star Trek*, see Michael (202–204) and Higley (153–4).

TOE as science's attempt to mythologize its own origins and maintain its power—to justify its goals and hence its requirements.

Literary critic Robert Markley has observed that "the efforts expended to defend the territories claimed by 'science' and the 'humanities' often seem an ironic measure of the instability of the distinction between them" (6). Big Bang cosmology is one of the focal points of this instability, with "science warriors," scientists, popularizers, philosophers of science and others wrestling over the boundary between scientific and mythic cosmology. Within this context, the popularizations examined in this chapter become more than straightforward explanations of scientific ideas. They make claims about the value and function of science in society, claims which impinge on cross-disciplinary relations. Each text begins by describing a myth—an essentially literary explanation of the world—and each ends with this explanation replaced by an entirely scientific one. Yet this process of enshrining the scientific worldview as the only meaningful one is itself a mythic process: both texts construct a teleological narrative in which science is the universe's ultimate product. "Science warriors" who criticize the constructivist blurring of myth and science thus need to look as much to their own as their opposition's tactics.

CHAPTER 6

Chaos, *Complexity* and Characterization: Stereotypes of the Scientist in Physics Popularizations

"Sitting at my desk or at some café table," writes Steven Weinberg in *Dreams of a Final Theory*, "I manipulate mathematical expressions and feel like Faust playing with his pentagrams before Mephistopheles arrives" (3). Weinberg's remark relies on the power of a particular fictional representation of the scientist to evoke a certain group of characteristics: when he mentions Faust, Weinberg bestows upon himself and his research connotations of magic, secrecy and danger. Faust is one of a number of ready-made scientist-stereotypes that constantly recur in books and films, and can also be conveniently drawn upon in purportedly factual representations of the scientist. These stereotypes, argues Roslynn Haynes in her book *From Faust to Strangelove: Representations of the Scientist in Western Literature* (1994), are a primary influence on the public image of the scientist: "Very few actual scientists (Isaac Newton, Marie Curie, and Albert Einstein are the only significant exceptions) have contributed to the popular image of 'the scientist'. On the other hand, fictional characters such as Dr. Faustus, Dr. Frankenstein, Dr. Moreau, Dr. Jekyll, Dr. Caligari, and Dr. Strangelove have been extremely influential …" (1). The scientific community is not oblivious to this: Patrick Parrinder notes that it was in the 1950s that "scientists in general became aware of the image-making power of science fiction," and the portrayal of scientists in all media became seen as a determining factor in the recruitment of young people into science (57). Popularizations of science are one source of images of the scientist—not so pervasive a source as popular films and fiction, but a source which has the added force of its claim to depict actual, rather than imaginary, science and scientists.

Yet, while fictional representations of the scientist receive much attention,[1] the ways in which the scientist figure is constructed by popularizers such as Weinberg have been largely ignored.[2] This is particularly surprising given that the later twentieth century saw a sharp decrease in the portrayal of scientist characters within science fiction (Parrinder 58), and an increase in visibility of scientist-popularizers. One reviewer of Haynes's book criticizes her exclusion of a number of sources outside

---

1 In addition to Haynes, see, for example, Parrinder; Weart; de Camp and Clareson; and Millhauser. There are also numerous analyses of scientist-figures in particular literary and film texts.

2 Exceptions include the recent analyses of Hawking's image listed in Chapter 5; and Davida Charney's article "Lone Geniuses in Popular Science."

of literature that are fundamental to the popular construction of the image of the scientist, giving as an example "Stephen Jay Gould's highly influential popular books" (Rousseau, "Virtual Realities" 229). It seems even stranger that a book entitled *From Faust to Strangelove* should fail to include any reference to Hawking, who, like Strangelove, is wheelchair-bound, and is the most visible scientist alive today.

The purpose of this chapter, therefore, is to make some preliminary investigations into the characterization of the scientist in physics popularizations, particularly the use of stock characters: stereotypes borrowed from fiction or elsewhere.[3] Like narrative and metaphor, characterization as a primarily literary concept is traditionally opposed to and excluded from scientific discourse, which is typified by what physicist David Mermin terms "the insistence on bland impersonality and the widespread indifference to anything like the display of a unique human author" (*Boojums* xi). The characterization of the scientist—his/her preconceptions, emotional state, career concerns—foregrounds exactly the factors that are not supposed to affect experimental results or theorizing. Popular science writing, by contrast, because it is designed to be popular, often reincorporates some form of "human interest." Yet personal details should not be dismissed as mere "colour" added to the scientific content. According to G. S. Rousseau, "Few topics today are more consequential for the funding of limited resources than the images held of scientists, technologists and medical professionals" ("Virtual Realities" 229). Popularizations of science can be an important means through which images of the contemporary scientist are constructed.

Popularizations, of course, vary significantly in the extent to which they focus on the figure of the scientist. At one extreme are texts in which "human interest" merely amounts to a list of names attached to theories or phenomena, rather than protagonists in the process of discovery. The actors that appear barely meet E. M. Forster's criteria for a "flat" character, let alone a round one. In popularizations of this kind, physical descriptions of scientists, their likes and dislikes, and particularly their feelings, tend to be excluded. The simple past is usually preferred to the continuous, emphasizing a discovery as an event with a name attached rather than an acted-out process. Direct speech is seldom used, free indirect speech almost never. Potted biographies are often included but provide little beside dates of birth and death, education, academic positions and Nobel Prizes. Consider, for example, John Gribbin's description of Chadwick: "James Chadwick, a British physicist who had been born in 1891 and was working with Rutherford ... confirmed, in a series of experiments in 1932, that neutrons exist; he received the Nobel Prize for his work, in 1935"; or Rutherford: "The New Zealander Ernest Rutherford, born in 1871, had established in 1910 that all the positive charge in an atom is concentrated in a tiny nucleus" (*In Search of the Big Bang* 263, 262). Granted, Chadwick and Rutherford are minor figures in Gribbin's narrative, and elsewhere he displays more interest in personality. Yet often, as here, he includes nationality and date of birth

---

3   I am not dealing here with scientific biographies. These texts constitute a genre of their own and cannot be grouped unproblematically with the type of popularizations I have looked at in previous chapters.

where no other personal information is given—it is as though, like science, personal description must be limited to falsifiable facts, not subjective guesswork. Certainly, there is a strong message in these popularizations that description of human experience and behaviour remains peripheral to the primary content of the text.

Other popularizers, however, sprinkle their texts with snippets of personal information about themselves or their colleagues, such as Weinberg's comment above. In the previous chapter I mentioned Hawking's references in *A Brief History of Time* to his own research; these often include an indication of his emotional state.[4] At several points Hawking also refers to his physical disability and his family life. In some cases, it is possible to see a development in the representation of the scientist throughout a particular popularizer's corpus. Paul Davies provides a good example: there is a chronological progression in his large body of works from textbook-style brevity to a far more anecdotal, intimate style. For instance, the man described in *The Forces of Nature* (1979) as "British physicist Paul Adrien Maurice Dirac" (74) becomes in *About Time* (1995) "the very model of an English middle-class gentleman, slightly stooped, with gray hair and mustache, and a quietly unassuming manner," whom Davies recalls having personally seen (143).

At the other end of the spectrum are popularizers who base their narrative around a particular scientist, making character one of the focal points of their exposition. Examples include Dava Sobel's bestseller *Longitude: The True Story of a Lone Genius Who Solved the Greatest Scientific Problem of His Time* and Simon Singh's *Fermat's Last Theorem*, both of which present the resolution of scientific and mathematical problems through a personalized narrative.[5] Other popularizations examine a network of scientists, but retain a focus on individual histories and personalities. A particularly successful example is James Gleick's 1987 bestseller *Chaos: Making a New Science*. Like the "new journalists" of the 1960s, Gleick borrows extensively from the techniques of the novelist to tell a non-fictional narrative. The reader is given insight into the protagonists' inner thoughts and feelings, provided with passages of dialogue, and presented with detailed descriptions of their physical appearance as well as the settings in which they live and work. This approach is even more noticeable in a slightly later popularization, M. Mitchell Waldrop's *Complexity: The Emerging Science at the Edge of Order and Chaos*, which shows remarkable similarity to *Chaos*, in its one-word title and longer subtitle, its style, its subject-matter,[6]

---

4 For example: "I must admit that in writing this paper I was motivated partly by *irritation* with Bekenstein, who, I felt, had misused my discovery ... I found, to my *surprise* and *annoyance*, that even nonrotating black holes should apparently create and emit particles at a steady rate ... I was *afraid* that if Bekenstein found out about it, he would use it as a further argument to support his ideas about the entropy of black holes, which I still did not *like*" (*A Brief History* 104–105; emphasis added; see also 49, 102, 116, 131).

5 See Davida Charney's "Lone Geniuses in Popular Science," in which she discusses Sobel's emphasis on the individual scientist, as opposed to group consensus.

6 Chaos theory and complexity theory are closely interconnected, and distinguishing between them can be difficult. Edward Lorenz notes that the two terms are often used interchangeably, and that both have many definitions (4, 163). Hayles suggests that chaos theory can be divided into "two main branches." The first branch, which includes researchers such as Benoit Mandelbrot and Mitchell Feigenbaum, and is described in *Chaos*, is "concerned

and its emphasis on character. Notably, all of these popularizations are authored by specialist science writers rather than scientists. As professional writers, these popularizers perhaps feel more at home with the techniques and approaches of the novelist than do those popularizers who primarily identify as scientists.

A striking aspect of the last two popularizations mentioned above, *Chaos* and *Complexity*, is that they borrow not only the standard devices of popular fiction, but also at times the more specific conventions of a well-established genre within popular fiction: hard-boiled detective fiction. In the first part of this chapter I argue that Gleick and Waldrop draw on this particular genre partly because the methods of the hard-boiled detective correspond to those of the scientists involved in chaos and complexity. But this stereotype carries with it more than a method: it also projects a certain set of personal characteristics upon the scientist. In particular, it implies that the scientist is isolated from normal social interaction and conventions.

The hard-boiled detective is only one of a series of stereotypes evident in prominent late twentieth-century popularizations that present the scientist figure as one at odds with social convention. These stereotypes are not identical to those that Haynes observes in fiction, but there is considerable overlap between the two sets. They include the scientist as "absent-minded professor," the scientist as priest or Zen-master, and the scientist as obsessive. All of these stereotypes are united by an emphasis on "outsider-dom": a sense that the scientist is in some way removed from everyday life. The extent to which the scientist (and science itself) is embedded in his/her society and culture has been one of the issues around which the "Science Wars" have been waged. Thus the way in which a popularizer chooses to represent the scientist can be read not simply in terms of its ability to generate reader interest, but also as a contribution to ongoing cross-disciplinary debates.

Characterization in *Chaos* and *Complexity*

Even a cursory glance at *Chaos* and *Complexity* is enough to reveal that these two physics popularizations are centrally concerned with character.[7] Gleick and

---

with the order hidden within chaotic systems." The second, represented by researchers such as Ilya Prigogine and Arthur Winfree, "focuses on the order that arises out of chaotic systems" ("Complex Dynamics" 12, 15). Complexity theory appears to have evolved from the second branch. Waldrop states that "complex systems" are those which "have somehow acquired the ability to bring order and chaos into a special kind of balance" (12). I have not attempted to define or distinguish further the two theories (which generally go under different, more specific labels in professional scientific discourse) as this is irrelevant to my concerns here, and is best gathered from the texts themselves.

7 The categorization of these two texts as "physics popularizations" is contestable. This difficulty stems from the fact that both research areas are inherently interdisciplinary. A central theme of Gleick's text is the way in which chaos crosses disciplinary boundaries, developing via the interactions of researchers in highly disparate fields (3–4). Waldrop's scientists are similarly eclectic; his central protagonist is an economist. Nevertheless, there is some justification for placing these texts within the broad category of physics popularizations. Gleick himself groups chaos together with two earlier "revolutions" in physics, relativity and quantum physics (6), and later states that "it was to be the physicists

Waldrop avoid the pedagogical style employed by many popularizers in which a single authoritative voice explains scientific concepts, preferring a more dynamic history of the researchers themselves. Gleick focuses on several central protagonists and a large number of bit-players, leaping from character to character in no clear chronological order. As the narrative continues, however, the links between the various characters are revealed, and the seemingly disparate strands of narrative are tied together to form a coherent history of the development of chaos theory. Waldrop's approach is similar, although not quite as fragmentary: he retains the unifying concept of one central character to whom he constantly returns, and whose career path provides the scaffolding for the text's narrative. Waldrop states explicitly in his prologue that his popularization focuses on researchers rather than research. After describing his protagonists and their theories, he ends his prologue with a dramatic four-word paragraph: "This is their story" (13). Both popularizers frequently provide seemingly irrelevant information about scientists' appearance, emotional state and personality: "John Hubbard, an American mathematician with a taste for fashionable bold shirts, had been teaching elementary calculus ..." (Gleick 217);[8] "Born and raised in New York City, his dark-rimmed glasses and white crew-cut hair giving him the look of a cherubic Henry Kissinger, Gell-Mann was brash, brilliant, charming, and incessantly verbal—not to mention being self-confident to the point of arrogance" (Waldrop 74).

Unlike most popularizers, but like most novelists, both Gleick and Waldrop make ample use of direct and indirect speech in order to give depth to the characters they describe. Waldrop constantly recreates incidents in the lives of researchers in a manner that, removed from its context, would be indistinguishable from that of the popular novelist:

> About a month after his ill-starred trip to Berkeley, as Brian Arthur was walking across the Stanford University campus on a sunny California day in April 1987, he was startled to see a bicycle pull up in front of him bearing a distinguished figure in sports coat, tie, and battered white bicycle helmet. "Brian," said Kenneth Arrow. "I was just going to call you."
> Arrow. Arthur was instantly on the alert. (52)

As this passage shows, Waldrop employs an omniscient narrative voice in *Complexity*. He has access to his characters' most intimate thoughts, and much of his text consists of free indirect speech similar to the following:

... who made a new science of chaos" (118). Waldrop's subject—the work of the scientists at the Santa Fe institute—is less easily categorized; however, a number of his main characters, such as Murray Gell-Mann, Philip Anderson and Doyne Farmer, are physicists. Roger Smith includes a section entitled "Chaos and Complexity" in his bibliography of popular physics books, noting that while these are "not exclusively theories of physics," it is true that "physicists have contributed to their development and ... they have potentially important ramifications for physics" (371); both Gleick's and Waldrop's texts are listed in the bibliography. Employing the same reasoning, I believe that it is appropriate to include these texts in my analysis, recognizing at the same time their drive towards interdisciplinarity.

8   All citations of Gleick in this chapter refer to *Chaos*, unless otherwise specified.

Oh yes, he'd really been high on himself that morning. So why hadn't he, years ago, just stuck to the mainstream instead of trying to invent a whole new approach to economics? Why hadn't he played it safe instead of trying to get in step with some nebulous, half-imaginary scientific revolution?

Because he couldn't get it out of his head, that's why. (16)

Gleick similarly has direct access to his characters' thoughts—"He did not know why he came to Santa Cruz" (244)—as well as their external interactions and dialogue: "William Burke, a Santa Cruz cosmologist and relativist, ran into his friend Edward A. Spiegel, an astrophysicist, at one o'clock in the morning in the lobby of a Boston hotel, where they were attending a conference on general relativity. 'Hey, I've just been listening to the Lorenz attractor,' Spiegel said" (244). Such uses of direct and free indirect speech create an effect of immediacy and intimacy between the scientist-characters and the reader.

This sense of intimate access to particular characters is amplified by other techniques employed by both popularizers, such as the use of the continuous past, which tends to emphasize the activity of the researcher, rather than simply the end result of the research: "Upstairs in the superconductivity laboratory, Shaw was making his desultory way to the end of his thesis work. But he was beginning to spend more and more time playing with the Systron-Donner" (Gleick 246); "Before the month of November was out he found himself walking near IIASA's Hapsburg palace, excitedly explaining increasing returns to a visiting Norwegian economist …" (Waldrop 38). The habitual past is used to similar effect: "'Nobody can say that I've rigged the system by choosing a pendulum,' Yorke would say jovially" (Gleick 234). Occasionally the present tense is employed to create a heightened sense of immediacy: "Mitchell Feigenbaum stands at streamside. He is sweating slightly …" (Gleick 157); or a character is introduced *in medias res*, in order to plunge the reader into the excitement of the story: "Years later, physicists would give wistful looks when they talked about Lorenz's paper on those equations …" (Gleick 30).

Each text features a number of dramatic, portentous chapter and section endings that focus as much on the activities of scientists as on their theories. Whereas a popularizer of cosmology or quantum physics might typically end a chapter by pointing to the mind-boggling implications of the theories described or about to be described, Waldrop's and Gleick's melodramatic announcements of impending scientific revolutions are frequently caught up with their characters' feelings and careers. Consider the first three chapter conclusions from *Complexity*: "Actually, he was right. And as it happens, he didn't have too much longer to wait" (51); "Arthur, he wrote on the list. Brian Arthur" (98); "Certainly that was true for Arthur. In his case, the discovery took about half a day" (143). Section endings from *Chaos* follow a similar pattern: "When he returned an hour later, he saw something unexpected, something that planted a seed for a new science" (16); "May, a man with one foot in each world, understood that he was entering a domain that was astonishing and profound" (77); "Ruelle added one piece of advice: 'Get in touch with Mandelbrot'" (216). Perhaps the most blatant example is the conclusion of Gleick's text:

> When he explained chemical chaos, displaying a transparency of a strange attractor, and said, "That's real data," a chill ran up Schaffer's spine.

"All of a sudden I knew that was my destiny," Schaffer said. He had a sabbatical year coming. He withdrew his application for National Science Foundation money and applied for a Guggenheim Fellowship. Up in the mountains, he knew, the ants changed with the season. Bees hovered and darted in a dynamical buzz. Clouds skidded across the sky. He could not work the old way any more. (317)

These observations show that Gleick and Waldrop use the techniques of popular fiction in order to construct character-centred popularizations. But a closer focus on the two texts indicates that they borrow from fiction more than general devices such as free indirect discourse. Both texts appropriate the style of a specific popular genre, the hard-boiled detective novel, constructing scientists in the mould of the tough, streetwise PI.

The Scientist as PI

The hard-boiled PI novel (or *roman noir*), as epitomized by the works of Dashiell Hammett and Raymond Chandler, is usually understood as a reaction against the perceived artificially of earlier detective fiction. It first evolved in the United States in the 1920s, developing during years of depression, prohibition, organized crime and high violence. Its protagonist is not a gentleman-amateur like Arthur Conan Doyle's Sherlock Holmes, or a maiden aunt like Agatha Christie's Miss Marple, but a professional Private Investigator (PI) who works for money. He (the typical PI is male) operates not in country villages and manor houses but in the rainy streets of Los Angeles or San Francisco. He deals with the "reality" of crime—gangsterism, gambling, pornography, corrupt cops—and uses his gun, his fists, his sarcasm and his knowledge of the streets and their inhabitants as much as his powers of deduction in order to catch his villains. The PI (or "gumshoe") is cynical, streetwise, and isolated: socially autonomous, he avoids emotional entanglements. He is tough and hard-living, drinking constantly and smoking countless cigarettes.

To any reader of hard-boiled PI fiction, the similarities between this genre and Gleick's and Waldrop's popularizations are immediately evident. Consider the opening paragraph of *Chaos*: "The police in the small town of Los Alamos, New Mexico, worried briefly in 1974 about a man seen prowling in the dark, night after night, the red glow of his cigarette floating along the back streets" (1). Cigarettes, nocturnal back streets, the prowling man, police: almost all of the atmospheric ingredients of the hard-boiled novel are present in this first paragraph. The opening of the first chapter of *Complexity* is even more striking:

> Sitting alone at his table by the bar, Brian Arthur stared out the front window of the tavern and did his best to ignore the young urban professionals drifting in to get an early start on Happy Hour. Outside, in the concrete canyons of the commercial district, the typical San Francisco fog was turning into a typical San Francisco drizzle. That was fine by him. (15)

Again, the main ingredients are present: a lonely scientist/detective in a drab cityscape, taking solace in alcohol, cut off from the society around him, cynically embracing bad weather. The setting is perfect, recalling the foggy San Francisco of Hammett's fiction.

The *noir* atmosphere established by these openings is reinforced by the conversational colloquialisms employed in each text. Colloquialisms are a staple way in which popularizers give their readers an impression of familiarity. However, the colloquialisms in *Chaos* and *Complexity* often have a toughness about them reminiscent of the hard-boiled PI: "'I understand you're real smart,' Agnew said to Feigenbaum. 'If you're so smart, why don't you just solve laser fusion?'" (Gleick 2); "It was the kind of talk that could make a sober physicist's head spin" (Gleick 262). Even Gleick's narratorial voice occasionally takes on the ironic, worldly-wise tones of Raymond Chandler's hero: "Later, people would say that James Yorke had discovered Lorenz and given the science of chaos its name. The second part was actually true" (65). And again, in Waldrop's text the parallels are even stronger:

> At Princeton, Philip Anderson got the note from Pines on June 29, 1984. Would he like to attend a workshop that fall on 'Emerging Syntheses' in science?
>
> Hmm. Maybe. Anderson was skeptical, to say the least. He'd heard rumors about this outfit. Gell-Mann had been talking up the institute everywhere he went ... .
>
> Well, Anderson could match credentials with Murray Gell-Mann any day of the week, thank you ... .
>
> Besides, he thought, this Santa Fe bunch sounded like a pack of amateurs ... . (79–80)

The streetwise colloquialisms—"this outfit," "talking up," "Santa Fe bunch," "pack of amateurs"—are straight from hard-boiled fiction. Compare this passage to the following dialogue from the screenplay of the classic *noir* film *Double Indemnity*, co-written by Chandler: "What kind of *outfit* is this anyway? Are we an insurance company, or a *bunch* of dimwit *amateurs*...?" (Wilder and Chandler 889; emphasis added). Other tough colloquialisms pepper Waldrop's text: "He just couldn't buy it" (23); "No, as the gambler once said, the game may be crooked, but it's the only game in town" (24); "Computers were nice, they felt ... But damn it, *another* computer research center? (72);[9] "But what the hell. At they rate they were going, they were in no position to say no to anybody. It was worth a shot" (91).

Physicist Philip Anderson, in the above passage, is portrayed as cynical and self-reliant, like the typical hard-boiled PI. A large amount of scholarship has been devoted to analysing this stereotype, which is generally understood as an attempt to embody an ideal of masculine autonomy and individualism. As novelist and critic Robert Parker observes, the hard-boiled detective "has no friends, no family, and no permanent social context" (4). Mitchell Feigenbaum, another physicist and one of the pivotal characters in *Chaos*, is constructed along very similar lines. He has little connection with domesticity or family life of any kind, and lives dysfunctionally. Like Sam Spade or Philip Marlowe, he is constantly found "puffing on a cigarette" (157); when absorbed in his work, he doesn't sleep or eat, presumably "getting his vitamins

---

9   "Nice" is an adjective particularly favoured by Chandler's PI Philip Marlowe: "Kissing is nice, but your father didn't hire me to sleep with you" (*The Big Sleep* 108). The word is used more than fifty times in *The Big Sleep*, mostly by Marlowe.

from cigarettes" (179). Gleick emphasizes his physical and emotional detachment from social networks. When, as a child, he "realized he could learn things," he became "more and more detached from his friends. Ordinary conversation could not hold his interest," and he was only able to re-enter social interaction through "a deliberate project" (159). He is constantly depicted alone and thinking: pacing the streets and thinking (1), listening to records and thinking (163), walking or standing apart from a group and thinking (163, 157). To complete the image, Feigenbaum is even indirectly connected with the guns so ubiquitous in hard-boiled detective fiction: "Like a gun collector wistfully recalling the Colt .45 in the era of automatic weaponry," writes Gleick, introducing a discussion of Feigenbaum's explorations on a Hewlett-Packard calculator, "the modern scientist nurses a certain nostalgia for the HP-65 hand-held calculator" (171).

These parallels offer convincing evidence that Gleick and Waldrop are (perhaps subconsciously) borrowing from the conventions of the hard-boiled detective novel. This is not, admittedly, the only source from which they borrow; both popularizations also draw from a range of other stereotypes, sometimes combining two or more stereotypes in the depiction of one scientist. Nor is the parallel sustained consistently throughout each text: the similarities to PI fiction are most obvious near the start of each. In both cases, however, the parallels that are constructed are strong enough to warrant an investigation into the possible reasons behind them. What work does the stereotype of the fictional hard-boiled PI do for these two physics popularizers? One possible answer is straightforward: by borrowing the well-established style and stock character of a highly successful subgenre of popular fiction, particularly in the early stages of the text, each popularizer exploits a formula which has proven success in generating reader interest. But any subgenre of popular fiction might have this effect; why is the detective novel chosen, and why the hard-boiled subgenre in particular?

The answer to the first part of this question lies in the parallels between the scientist, as someone searching for the clue to a particular puzzle in the physical world, and the detective, who attempts to solve a different kind of puzzle. The scientist-as-detective is a well-recognized fictional stereotype.[10] Most often, however, the parallel with the scientist is assumed to be what is sometimes termed the "classical" detective—the detached, deductive genius who solves crimes primarily through his or her powers of reason—rather than the PI. Haynes places the scientist-as-detective stereotype within her wider "scientist as hero" category, positing Sherlock Holmes as its "prototype" (178). Doyle establishes the link between Holmes and the scientist early in the inaugural Holmes story *A Study in Scarlet*: Watson and the reader first sight the detective over tables of test-tubes and Bunsen burners, as he cries aloud, Archimedes-style, "I've found it! I've found it!" (13); the scientific nature of his method is expounded in the second chapter, entitled "The Science of Deduction." The continuing popularity of Holmes in particular as an embodiment of scientific rationality is reinforced by Colin Bruce's popularization *The Strange Case of Mrs Hudson's Cat Or Sherlock Holmes Solves the Einstein Mysteries* (1998), in

---

10 See, for example, Haynes 178–9; Curtis; Agassi; Hudson; and P. Rose.

which Holmes, Watson and several other of Doyle's staple characters grapple with problems from classical physics, relativity and quantum theory.

Sherlock Holmes, however, is only one kind of detective. His character and method are suited to explications of certain kinds of science—in particular, sciences that prioritize (or appear to prioritize) abstract theorization. Literary critic Stephen Knight notes that Holmes's claim that his method is both deductive and scientific is inconsistent: "if Holmes really were finding patterns in facts he would be practising 'induction'" (86). Rather, Holmes practises only the deductive side of the scientific method—"drawing from a set of existent theories to explain new events" (86). For Knight, this confusion of the scientific method with pure deduction reflects the conservatism and need for social stability of Doyle and his audience: "The dress of modern materialist science is used for conservative thinking, for a failure to face the real, disorderly experience of data" (86). Phyllis Rose in her article "Huxley, Holmes and the Scientist as Aesthete" argues that while Doyle describes Holmes's method of deduction in much the same way as T. H. Huxley described the scientific method, he nevertheless portrays Holmes in opposition to Huxley's "common sense"-based approach—as an elite, arrogant figure, whose powers of deduction are unique and unapproachable. She convincingly likens Holmes, with his cocaine habit and long spells of withdrawal from social interaction, to the decadent aesthete: he "views logic as an aesthetic process ... [he] is an artist" (23–4). Both of these descriptions of Holmes suggest a parallel with the figure of the cosmologist or particle physicist. The previous chapter discussed the emphasis placed on theoretical deduction and its aesthetics—the search for "elegance" and "beauty"— rather than empirical investigation in these fields, a trend which some critics believe began (in the modern instance) with Einstein. It is unsurprising then, that in their popularization *The Evolution of Physics*, Einstein and co-author Leopold Infeld call on the classical detective stereotype, noting that "[i]n nearly every detective novel since the admirable stories of Conan Doyle" the investigator must do some "pure thinking," retiring Holmes-style to dwell on the facts—facts which "often seem quite strange, incoherent, and wholly unrelated"—already collected: "So he plays his violin, or lounges in his armchair enjoying a pipe, when suddenly, by Jove, he has it!" (4). For similar reasons, Colin Bruce, in his popularization of quantum theory and relativity, chooses Holmes, the master of deduction, as his model investigator. Although Bruce states in his "Afterword" that "[i]t is a capital mistake to theorize ahead of the facts" (251), within the text there are a number of occasions where deductive reason is emphasized over empiricism (26, 27–8, 31–2).

The science that Gleick and Waldrop aim to explicate is very different. Both popularizers emphasize that they are popularizing "new" or "emerging" kinds of science. Chaos and complexity, as portrayed by Gleick and Waldrop, represent a move away from speculative, overly deductive reasoning to a method more suited to mundane, real-world situations. Gleick observes in his prologue that the scientists who developed chaos felt that physics "had been dominated long enough ... by the glittering abstractions of high-energy particles and quantum mechanics" (6). As chaos theory progresses, "the best physicists find themselves returning without embarrassment to phenomena on a human scale" (7). Waldrop's researchers also reject the methods of the previous "glamour" areas of physics. One physicist, for example, is "outraged at the way Congress lavished money on shiny new telescopes and fantastically expensive new accelerators while small-scale projects ... were starving" (80). More generally, Waldrop's characters turn away from a reliance on "precise, deductive logic" (253). The Santa Fe Institute enables its members to realize, "'Yes! We can deal with inductive rather than deductive logic'" (235). Induction, according to Waldrop, "is what allows us to survive in a messy, unpredictable, and often incomprehensible world" (253). In Waldrop's representation, inductivism involves not simply methodical collection and collation of data, but rather an eclectic group of pragmatic approaches to problems. Inductivists, he explains, "try to fill in the gaps on the fly by forming hypotheses, by making analogies, by drawing from past experience, by using heuristic rules of thumb. Whatever works, works—even if they don't understand why" (253). In both popularizations, there is a sense that the "new" sciences of chaos and complexity represent a rejection of detached, aesthetic deductionism.

Moreover, in both popularizations, the sense that chaos and complexity work differently from earlier kinds of physics is evident in their form as well as their content. As discussed in Chapter 5, the typical popularization of cosmology is structured linearly: the narrative moves chronologically from one discovery to the next, or thematically from simple to more complex concepts. The impression given is one of smooth inevitable progress, which parallels both the inevitability of the cosmologist's deductive logic and the evolution of the universe. The same set of parallels between the phenomena being studied, the scientist's method and the narrative structure is evident in Gleick's and Waldrop's texts. The sciences of chaos and complexity develop in a haphazard, emergent manner that parallels both the development of chaotic and complex systems themselves and the inductive, ad hoc methods of chaoticians (as presented by these popularizers). Gleick's narrative appears to jump arbitrarily from one researcher to another, jerking backwards and forwards in time and space. The list of chapter contents emphasizes this disjointedness; for example, Chapter 4, "A Geometry of Nature," is given the following description: "A discovery about cotton prices. A refugee from Bourbaki. Transmission errors and jagged shores. New dimensions. The monsters of fractal geometry. Quakes in the schizosphere. From clouds to blood vessels. The trash cans of science. 'To see the world in a grain of sand'" (ix). Only gradually, as the characters begin to interact, does a pattern—an outline of a "new science"—begin to emerge. Chaos theory does not, in Gleick's narrative, evolve smoothly, linearly or inevitably; rather, like the systems it describes, it displays sensitive dependence on initial conditions. Random links between Gleick's researchers—a chance meeting, or an accidental series of events—lead to important exchanges of ideas. Just as the science of chaos theorizes the way in which "the disorderly behavior of simple systems [acts] as a *creative* process" (43), so in *Chaos*, from a group of researchers working in seemingly disparate fields, interacting in an unstructured, seemingly random way, emerges the coherent body of knowledge that is chaos theory. A similar analysis could apply to Waldrop's text, in which a disparate group of researchers merge to form the Santa Fe Institute without any apparent guidance from a unifying body. Waldrop himself identifies this analogy: "The joke among the science board members was that the Santa Fe Institute was an emergent phenomenon all by itself.

It was a joke they actually took quite seriously" (248). The narratives of both *Chaos* and *Complexity* are clearly structured to parallel the phenomena they describe; the researchers act like components of a chaotic system, working in a haphazard yet ultimately highly productive manner.

The figure of the aesthetic, deductive Holmesian detective deployed by Einstein and others would clearly be out of place in popularizations that depict researchers as operating in a far more spontaneous, ad hoc way. In an analysis of popular science magazine articles that adopt the detective narrative format, Ron Curtis argues that it is the "hard-working scientific 'gumshoe'" rather than the "vain, arrogant, disputatious cosmologist" who is the "modern counterpart" of the ideal inductivist scientist envisaged by Francis Bacon (422, 444). To understand further why the PI or "gumshoe" is a particularly useful stereotype for Gleick and Waldrop, it is helpful to explicate the differences between this character and that of the classical detective. T. J. Binyon in his *'Murder Will Out': The Detective in Fiction* observes that these two figures, while "similar superficially," are actually "radically opposed to one another" (32). He notes that the classical detective operates in a "[c]losed society, with limited number of suspects, who are introduced at the beginning of the narrative," while the PI operates in an "[o]pen society, with indefinite number of suspects, who are introduced throughout the narrative." The classical detective is "usually hired to solve a crime," whereas the PI is "usually hired to investigate a situation." The classical detective is "basically static: remains in one place to interview suspects," whereas the PI is "basically mobile; moves from place to place to interview characters" (32). More fundamentally, one could contrast Holmes's careful logical deduction with the hard-boiled PI's haphazard trailing of clues: the latter relies more on being in the right place, at the right time, with the right person, than on intellectual reasoning. As Joseph Agassi observes, Chandler's hero is happy to chase "accidental red herrings" and "clues leading nowhere" (104). Hammett's Continental Op, when confronted by another character with the accusation that "you scientific detectives" have "the vaguest way of doing things I ever heard of," observes, "Plans are all right sometimes … And sometimes just stirring things up is all right—if you're tough enough to survive, and keep your eyes open so you'll see what you want when it comes to the top" (79).

For Gleick and Waldrop, who are eager to portray the disorderly nature of both scientific progress and physical systems, the method—or anti-method—of the PI thus provides a far more suitable framework than the reflective, deductive approach of Holmes and his successors. "Stirring things up": Hammett's description of the PI's haphazard method provides an ideal metaphor for chaos theory on both a physical and an epistemological level. Chaoticians investigate phenomena such as turbulence, but they themselves are seen as adding turbulence to the totalizing, reductionist approaches of cosmology and particle physics. As Gleick observes, "some of those who thought physics might be working its way into a corner now look to chaos as a way out." This escape is via a return to familiar objects and experiences: as mentioned above, Gleick presents chaos as a science of the everyday, of human-sized objects rather than the very big or very small (6). This fits nicely with the hard-boiled PI's willingness to perform mundane or routine tasks, and to submerge himself in the sordid world of his client rather than retire to contemplation.

Like the PI, the founders of chaos theory do not set out to solve a great mystery, but rather to explore a situation without knowing quite what they are looking for. Whereas both quantum theory and relativity developed from crucial paradoxes in existing theories, chaos theory, as Gleick tells it, was stumbled upon by researchers when computers became sufficiently advanced to make this stumbling possible. Edward Lorenz's breakthrough comes when he takes "a shortcut" with the data he enters into his computer, leaving out some decimal points (16); Gleick informs us unequivocally that "Lorenz's discovery was an accident" (21). Christopher Scholz "*stumbled* across Benoit Mandelbrot's name" as a graduate student (103; emphasis added), remembered it later, bought Mandelbrot's book, and became known in his field for using fractal techniques (104–106). Several of Gleick's scientists find their way to chaos theory while investigating a seemingly unrelated area (May was a physicist studying population ecology, Mandelbrot a mathematician at IBM working on noise in telephone lines), or pursue their interest in it to the detriment of their official task (this is true, for example, of Feigenbaum [2–3] and Shaw [246–9]). These researchers make discoveries when they least expect it, find chaotic systems while they are investigating a different phenomenon, obtain a vital insight from a seemingly arbitrary or unrelated meeting with another character. Waldrop's researchers similarly have no clear "plan," but, like the PI, they progress by "stirring ... up" existing assumptions and seeing what "comes to the top." Their method is the haphazard, mobile, involved and "messy" method of the PI, rather than the detached deduction of the classical detective.

There is one way, however, in which the stereotype of the PI differs relatively little from that of the classical detective. Both of these fictional types are marked by their distance from social networks. Haynes writes, "Holmes is characterized not only by a dedication to analysis and objectivity but also by his coolness and lack of social involvement" (178). Stephen Knight observes that although Holmes does have some "contact with normality," sharing his comfortable home with Watson and enjoying his engagement with his clients, he is also simultaneously "a certain distance from human normality" (80, 79). Emotional involvement, Knight argues, would "damage the scientific force" of Holmes's investigations; his "aloof, sometimes arrogant personal qualities" mesh with the "dispassionate isolation arising from [his] scientific powers" (79). Chandler takes the autonomy of the detective even further. He "isolate[s] the hero by making his clients untrustworthy," and realizes "a thorough theory and practice of the alienated individual defending himself against threats of the external world" (Knight 92, 163). Despite Gleick's interest in presenting chaos theory as the product of an interacting network of scientists rather than a single genius, he nonetheless carries over this aspect of the hard-boiled detective stereotype into his portrayal of chaoticians. Similarly Waldrop, while emphasizing the way in which complexity emerged out of a group of scientists interacting, opens his text with an image of a lonely researcher cut off from others. While the PI stereotype helps both popularizers to convey the new methods of chaoticians, it simultaneously perpetuates a long-standing stereotype of the socially isolated scientist.

Thus, where Gleick's and Waldrop's hard-boiled chaoticians might differ in their methods from Holmesian cosmologists and quantum physicists, both stereotypes converge in their portrayal of the scientist as removed from the surrounding social world. This is unsurprising, because many of the various stereotypes adopted by physics popularizers share this same quality. I want now to briefly outline some of the other stereotypes that appear in contemporary popularizations (including Gleick's), before looking more abstractly at recent debates about the relationship between science, the scientist and society.

The Scientist as Absent-Minded Professor

One of the earliest "recurrent stereotypes" of the scientist identified by Haynes is that of the "stupid virtuouso" who is "out of touch with the real world of social intercourse": "preoccupied with the trivialities of his private world of science, he ignores his social responsibilities" (3). According to Haynes, this seventeenth-century stereotype finds its modern descendant in the "absent-minded professor" image (3).[11] In both cases, the independence of the scientist from culture is represented by his ignorance of everyday concerns and social etiquette. It is this stereotype that informs the popular cliché of the badly dressed, ill-coordinated scientist. As physicist Sydney Perkowitz notes in a *New Scientist* article entitled "Real Physicists Don't Wear Ties,"

> It is easy, and certainly soothing to the ego, for scientists to conclude that brilliant people searching for eternal truths need not bother with fashion's frivolities. It is not our dress, we believe, but our dedicated brainpower that leaves our mark on the Universe. Did Einstein or his admirers care in the least that the body housing this great mind wore a baggy sweater, sockless shoes, and ebullient hair? ... The image of the almost alarmingly incisive intelligence looking outward to nature's secrets, not inward to mere details like correct clothing and haircut, is part of our mythology. (23)

A search through physics popularizations will not reveal many examples of the much-loved cinematic image of the absent-minded professor as a doddery old man with white fly-away hair *á la* Einstein. This stereotype suggests, Haynes notes, an "ineffectual figure," and given that most physics popularizers emphasize the importance and excitement of the discoveries they report, such an image would not further their (or physics's) ends. Nonetheless, at least one prominent popularizer has been able to reshape this stereotype for his own purposes.

Richard Feynman created and sustained a powerful variation on the "absent-minded professor" stereotype. Gleick in his biography of Feynman describes his subject's tendency to mythologize his own history through the repeated telling of anecdotes, many of which are collected in his two bestselling popularization-cum-memoirs *"Surely You're Joking, Mr Feynman!"* and *"What Do You Care What Other People Think?"* In these stories Feynman brings together a variety of stereotypes to create his image: according to Gleick, he was "a gadfly, a rake, a clown, and a naïf" (*Genius*

---

11 Bernadette Bensaude-Vincent claims that this stereotype can be traced back to Socrates: "the image of the absent-minded scientist ... is as old as western science" ("A Genealogy" 100).

11). This last stereotype, the naïf, encapsulates the same happy obliviousness to social etiquette that characterizes the absent-minded professor image. In *"Surely You're Joking,"* Feynman constantly depicts himself standing outside the social world: arriving at Princeton University with "no social abilities whatsoever," the young Feynman asks his hostess for cream *and* lemon in his tea, and receives the response which provides the title for his collection (60); not understanding "what exactly it meant to be 'social,'" he asks a waitress to a fraternity dance and is quickly found a more appropriate date by his classmates (31); sitting in on an advanced course in cell physiology, he needs some information on the physiology of the cat, so asks the shocked biology librarian for "a map of the cat" (72).

Behind all these anecdotes is the image of the wise fool; or, as Gleick puts it, "the boy who saw the emperor with no clothes" (*Genius* 11). Feynman's lack of knowledge of the niceties of each situation is always shown to result in a debunking of unnecessary or damaging practices. The hostess who laughs condescendingly is trapped in a trivial and snobbish etiquette against which Feynman's honest bumbling appears refreshing. The "social guys" who teach Feynman to meet girls and then reject his waitress date display a similar elitism (31). Feynman's effort in the library may have started rumours about "some dumb graduate biology student," but the biology students who interrupt Feynman during his talk on the physiology of the cat to tell him they already know the information are the real dummies: "They had wasted all their time memorizing stuff like that, when it could be looked up in fifteen minutes"; Feynman can catch up their four years of learning easily (13). Feynman displays the same unmasking innocence during his encounters with more sinister forces such as military conscription, anti-Semitism and the Manhattan Project.

This social naïf stereotype thus works on two levels: it fulfils the same comic function as the absent-minded professor image, and it reinforces the scientist's ability to step outside preconceived notions. In Feynman's anecdotes there is constant slippage between social and intellectual conventions; both come to represent exactly those forces that the brilliant scientist can and must ignore. It is as though the discarding of social conventions is equivalent to the discarding of unscientific attributes.

The Scientist as Priest or Zen-Master

Where the absent-minded professor stereotype assumes a happy oblivion to social conventions, another stock image of the scientist reinforced by physics popularization suggests a repudiation of everyday, material concerns. Physicists, due to the esoteric and seemingly elitist nature of their work, are often attributed priest-like qualities; they have, Brian Appleyard complains, become "a new priesthood" ("God and the Scientists"). In *Pythagoras' Trousers: God, Physics, and the Gender Wars*, Margaret Wertheim explores the historical connection between physics and the concept of a masculine priesthood. She suggests that the later twentieth century saw a particular emphasis on the priestly role. This was due partly to the legacy of Einstein, who "put the transcendence back into science. After a century during which physics had become increasingly concerned with the prosaic and the practical, Einstein once again turned physicists' gaze heavenward" (176). Wertheim argues that Einstein's "cosmological theory, his eminently quotable remarks about God, and his enigmatic statements about the process of science itself" have become "woven together to create a public persona of the physicist as religious mystic—an image he was the first to encourage" (186). But the prevalence of the scientist-as-priest image in recent decades is also, according to Wertheim, a result of an increasing disenchantment with traditional religion: "in an age when many people are hungering for a rapprochement between the spiritual and the scientific, the concept of the physicist as high priest is immensely appealing" (218).

While popularizers do not deploy this stereotype in the concrete ways that Feynman appropriates the "absent-minded professor" image, their tendency to turn, at the end of their texts if nowhere else, to questions usually dealt with by theologians reinforces the sense that they are a secular priesthood. As discussed in Chapter 1, popularizations that deal with the relationship between physics and "God" were very successful during the boom years. And even when no explicit mention is made of God or religion, popularizers often adopt an unacknowledged mythic rhetoric. The association of physics and the divine established by Einstein has, according to Wertheim, "been taken up with a vengeance by his successors, foremost among them Stephen Hawking" (176). The previous chapter looked at Hawking's public image in terms of his own mythic narrative in *A Brief History of Time*, noting that he textually positions science as the ultimate goal of the universe, and himself as the latest in a series of scientific forefathers. This self-mythologization is aided by Hawking's appearance: his physical immobilization gives the sense that he is a mind divorced from the needs of the body. In this sense he becomes an extreme version of the priest or mystic who eschews material, bodily existence in order to achieve transcendence in another realm.

Another important version of the scientist-as-priest stereotype is the "Zen-master" image constructed by the proponents of quantum mysticism such as Fritjof Capra and Gary Zukav. Andrew Ross has identified some obvious ways in which Capra constructs the physicist as an Eastern mystic: "Capra expressly points to the long years of training shared by the apprentice physicist and the Zen student in pursuit of their highly specialized crafts. Both are initiated into serious enquiry about the nature of the universe, and both become the curators of advanced knowledge about secrets that are inaccessible to the uninitiated" (*Strange Weather* 44). Zukav is even more blatant in his conflation of the physicist and the mystic. As noted in Chapter 4, there is a constant—and conscious—slippage in his text surrounding the concept of Wu Li. Wu Li ("Patterns of Organic Energy") represents subatomic reality, the "dance" of particles/waves. The Wu Li Masters, the physicists who investigate this reality, also perform a dance of energy: "The Wu Li Masters move in the midst of all this, now dancing this way, now that, sometimes with a heavy beat, sometimes with lightness and grace, ever flowing freely" (43); physicists are "dancing with Kali, the Divine Mother of Hindu mythology" (330). Through such metaphors Zukav portrays his physicists as creatures mystically aligned with the dynamic movement of the universe. Furthermore, Zukav emphasizes that not every physicist is a Wu Li Master (35). He is eager to separate the "'scientist,'" who "seeks to know the true nature of reality" and "deals with the unknown," from the "'technician,'" who is "a highly trained person whose job is to apply known techniques and principles." Most "'scientists,'" Zukav informs his reader, are in fact technicians (36). Zukav's book is "about Wu Li Masters and not about technicians"; thus he tells his reader, "we will use the word 'physicist' from now on to mean those physicists who are also scientists, that is, those physicists (people) who are not confined by the 'known'. From the little that we know about Wu Li Masters, it is evident that they come from this group" (42).

What emerges from Zukav's highly metaphorical language is a very selective representation of the activities of physicists which, as in Capra's book, makes them appear even more elitist and esoteric than previously imagined. Wu Li Masters "perceive in both ways, the rational and the irrational, the assertive and the receptive, the masculine and the feminine. They reject neither one nor the other. They only dance" (65). What is excluded from Zukav's picture is the often monotonous, day-to-day work in a laboratory or at a computer terminal that is the lot of almost all physicists (even those who "seek the true nature of reality"). All the social elements of science which science studies practitioners are currently bringing to light are gone; the physicist as pictured by Zukav is as far from the mess of experiments, departmental politics and arguments over authorship as Buddha under his banyan tree.

The Scientist as Obsessive

Closely related to the image of the scientist as priest or Zen-master is that of the scientist as obsessive—someone who is so absorbed in his[12] work that all other concerns become trivial. This stereotype can be classed under Haynes's category of "the alchemist, who reappears at critical times as the obsessed or maniacal scientist" (3); her central example is the figure of Faust. The scientist-as-obsessive is the glamorous, slightly dangerous equivalent of the absent-minded professor: where the latter is vague and comic, the former is driven and (potentially) tragic. And like the absent-minded professor stereotype, the seemingly negative image of the scientist-as-obsessive can be appropriated by popularizers to more positive ends. As Haynes notes, the Faust character in literature is ambiguous: the earliest written version of Faust's story is highly moralistic and emphasizes his presumption, but Christopher Marlowe's early seventeenth-century play the *Tragical History of Doctor Faustus* implicitly bestows a sense of the tragic hero on the character, while Romantic re-tellings (such as Goethe's) emphasize the nobility of the Faust figure in attempting to "transcend the limitations of the human condition" (Haynes 19). When, in the quotation that opens this chapter, Steven Weinberg likens himself to Faust (it is Goethe's play to which he directly alludes), he is inviting the reader to view him through a stereotype associated with secrecy, excitement, glamour and a noble desire for knowledge. Weinberg translates this stereotype into a modern setting: he manipulates mathematical expressions while at his desk or "some café

---

12 As I argue below, this and the other stereotypes described here are gendered: they assume a male physicist.

table." He is, it seems, careless of where he works or which café he frequents; his research is enthralling enough to follow him to the café table, and so abstract and removed from laboratories or equipment that it can be pursued in such an environment. There is a suggestion here of the obsessive scientist who is blind to everything but the pursuit of his goal.

While Weinberg only hints at a sense of obsession, other popularizers take this stereotype further. Gleick, for example, deploys it in *Chaos* to describe several of the scientists involved in this field. Earlier I mentioned the character of Mitchell Feigenbaum, whose chain-smoking and social isolation link him to the stereotype of the hard-boiled PI. But Gleick's portrayal of Feigenbaum's obsessive nature at times goes beyond this stereotype to tap into an older Faustian image. Feigenbaum is initially observed "experimenting with twenty-six-hour days," deliberately removing himself from regular rhythms, oblivious to trivialities such as the sun rising or setting (1). He lives in "a bare space, a bed in one room, a computer in another ..." (184). Once, the reader is told, when he was preoccupied with one particular idea, Feigenbaum worked non-stop for two months "as if in a trance," hardly sleeping. Even in peaceful periods, he subsists on the vaguely demonic diet of "the reddest possible meat, coffee, and red wine" (179). His physical description is given on the second page of *Chaos*:

> His hair was a ragged mane, sweeping back from his wide brow in the style of busts of German composers. His eyes were sudden and passionate. When he spoke, always rapidly, he tended to drop articles and pronouns in a vaguely middle European way, even though he was a native of Brooklyn. When he worked, he worked obsessively. (2)

Instead of the fly-away hair of the absent-minded professor, Feigenbaum has the "ragged mane" of the maniacal, obsessive scientist. In his single-minded pursuit of intellectual understanding, he discards trivial details such as parts of speech. Later we learn that Feigenbaum reads Goethe and "revel[s]" in Faust, "soaking up" its combination of passion and the intellect (163). This stereotype of the obsessive, like the others discussed above, is one in which the scientist is detached from social networks and heedless of the habits and conventions (such as pronouns, furniture and twenty-four hour days) which regulate the lives of others.

"Innocents and Outsiders"

In order to understand how these character stereotypes play a part in interdisciplinary debate, it is necessary to outline recent arguments about the relationship between science, individual scientists and the society in which both are situated. The representation of science in the public arena, according to science communication researcher John Durant, is often marred by the simplistic projection of the attributes of science upon scientists themselves:

> The most serious weakness in the standard view of the processes of scientific inquiry is its tendency to project the qualities of scientific knowledge upon the individual scientists who produce it. ... The projection of the characteristics of science upon its practitioners

is partly responsible for the public image of scientists as super-men and super-women; but this projection obscures the true nature of science and makes it all the more difficult to understand the course of science in public. ("What is Scientific Literacy?' 135)

Durant recounts occasions, such as the cold fusion scandal, when individual scientists publicized their findings without first receiving the endorsement of the scientific community—findings that turned out to be unwarranted. In the confusion that ensued, the general public, accustomed to viewing scientists as super-people, observed the events with bewilderment (135). Durant concludes, "The public needs to understand that sometimes science works not because of but in spite of the individuals who are involved in the process of knowledge production and dissemination" (136). The gist of his argument seems to be as follows: if the public mistakenly equates individual scientists with science, and the individual scientist is subsequently seen to be fraudulent or incorrect, then the public loses faith in science itself.

A similar desire to acknowledge the humanity and hence fallibility of individual scientists was expressed by prominent scientist-popularizers in the 1990s, at the height of the popular science boom. Physicist Freeman Dyson, at a 1992 meeting of philosophers, scientists and fellow popularizers, stated that

> The image of noble and virtuous dedication to truth, the image that scientists have traditionally presented to the public, is no longer credible. The public, having found out that the traditional image of the scientist as a secular saint is false, has gone to the opposite extreme and imagines us to be irresponsible devils playing with human lives … . It is our task now to dispel these fantasies with facts, showing to the public that scientists are neither saints nor devils but human beings sharing the common weaknesses of our species. ("The Scientist as Rebel" 9, 11)

In 1997, in a two-page spread in the *Independent*'s "Sunday Review," embryologist Lewis Wolpert and his co-author television producer Alison Richards similarly complained that

> If [scientists] feature in popular awareness at all, it is through a limited set of media stereotypes. With a few exceptions, if scientists are not mad or bad, they are personality-free, their measured tones and formal reports implying ways of thinking and working far removed from the intellectual and emotional messiness of other human activities. ("The Insiders' Story" 44)

In their article, and the book from which it is taken, *Passionate Minds: The Inner World of Scientists*, Wolpert and Richards aim to "redress the balance, to give a rare glimpse of the human reality of scientific life" (44). As in the examples from Durant and Dyson, there is an emphasis on the human aspects of scientific activity, the emotion, the "messiness."

All of these commentators, however, deny that the humanness or "messiness" of science which they emphasize significantly influences the results of science in a constructivist sense. For Durant, science works "in spite of" individuality. Dyson insists that, "There is no necessary contradiction between the transcendence of science and the realities of social history. One may believe that in science nature will ultimately have the last word, and still recognize an enormous role for human vainglory and viciousness in the practice of science before the last word is spoken" (10). He thus does not "fully agree" with social historians who believe that "science is driven by social forces" (9). Instead, he argues that scientists should be viewed as "free spirits ... rebelling against the local tyranny that each culture imposes on its children" (2). Wolpert, whose dismissal of relativist-constructivist approaches I discussed in Chapter 3, is similarly sceptical of attempts to tie scientists' research too closely to their social circumstances. In *Passionate Minds*, he and Richards emphasize the benefits that result when scientists are immune to social and intellectual constraints. They argue that "being an outsider is a definite help when it comes to making important scientific discoveries," as it brings with it a "fundamental advantage—notably the ability to view a subject or problem with an eye unclouded by history, habit or dogma." "Innocents and outsiders," they claim, "may also disregard—or fail to register—conventional barriers" (2).

This view of the scientist as slightly at odds with, or removed from, social conventions is one that gels with other accounts of genius in a variety of fields, such as that put forward in Howard Gardner's *Creating Minds*. Gardner argues that creative geniuses such as Einstein, T. S. Eliot and Pablo Picasso shared an ability to access "the sensibility of the very young child" (8)—an image that corresponds to that of the "innocent" or social naïf. Gardner also attributes to the creators he studies qualities similar to that of the "obsessive scientist" discussed above: they all "became embedded in some kind of bargain, deal, or Faustian arrangement" in which they sacrificed any possibility of "rounded personal existence" to their creative "mission." This bargain might take the form of "self-imposed isolation" or dysfunctional relations with others (44). In each case, the creator experienced a "feeling of marginality" from the surrounding community which he or she exploited in the creative process (260). Viewed in this context, the above argument merely echoes larger arguments about the nature of creativity.

However, in the more immediate context of the "Science Wars," the promotion of the scientist as an outsider-figure is controversial, as it seems designed to combat science studies researchers' increasing emphasis on the extent to which scientists are influenced by, and science itself produced by, the particular society in which they operate. Science studies scholars are just as eager as Wolpert, Dyson and Durant to emphasize the "messiness" of scientific practice. Steve Woolgar, for example, claims that the sociology of scientific knowledge effects a "repopulation of scientific stories. In these revelations of the messy, idiosyncratic face of science, as carried out at the laboratory bench, the scientists themselves come back into the picture" ("What is a Scientific Author?" 182). However, these critics take a more sceptical view of "the ability to view a subject or problem with an eye unclouded by history, habit or dogma." They are less concerned about the bestowal of the characteristics of science upon scientists than about the converse process, the separation of science from those who practise it. Woolgar notes that in traditional sociology of science "the actual character of science (in particular, the esoteric details of the content of scientific knowledge) is treated as independent of (prior to) and separate from its practitioners" (*Science* 25). One of the purposes of the sociology of scientific knowledge is to challenge this division. In the anthology

The Literature of Science, mathematician David Stone similarly argues that if a "rapprochement" between "scientific and popular cultures" is to be reached, "The first step would be to recognize that the mythologized history of science, with its distinction between science in the abstract and what it is that scientists do, is a fiction whose fundamental purpose is to establish a voice of authority for science." One of the ways in which this is achieved, Stone claims, is "by reifying science as something above the fallibility of individual scientists." Scientists, he insists, need to recognize that "there is no science independent of actual scientists" (304).

There seems to be another paradoxical twist on the "two cultures" division at work here: scientists such as Wolpert and Dyson emphasize the "human" aspects of their activity because they want to distance the image of the scientist from science itself; constructivist sociologists and others also emphasize science as a human activity, but because they want to close the perceived gap between scientists and the science they produce. Popularizers become part of this debate when they depict characters who, like the absent-minded professor, fail to register conventional barriers, or, like classical and hard-boiled detectives, disregard them. All of these stereotypes stress the isolation of the scientist from the social mainstream—his distance from the society in which he works. They are ways of reintroducing detailed individual characters into popularizations while downplaying the specific cultural and social circumstances of those characters. They bolster arguments about the immunity of the scientist to social and cultural assumptions and conventions, rather than his situatedness in them. By adopting these stereotypes, recent popularizers (consciously or not) play a role in the "Science Wars."

Stereotypes and the Gendered Scientist

As my choice of pronoun in the above paragraph indicates, one corollary of this kind of representation of the scientist is the automatic exclusion of women from physics popularizations. The absence of women both as authors of and actors in popular physics books should be clear to anyone who has read many of these texts. To give an indication of the extent of this absence, I have used the simple measure of counting the number of names indexed in the five core texts discussed in this book, and noting how many are identifiably those of women.[13] *Chaos* lists about 150 men, and two women: one, a mathematician, is mentioned only in an endnote, and in conjunction with a male co-researcher (325); the other, biologist Lynn Margulis, is mentioned once in conjunction with James Lovelock as a proponent of the Gaia hypothesis (278). *Complexity* also lists over 150 men, and twelve identifiable women: four of these are wives of male researchers; those remaining consist of a provost at a university; the director of programs at the Santa Fe Institute, whose qualifications are unspecified; an economist; an unspecified academic—presumably an economist; a bank employee with a statistics background who assists a businessman; a computer scientist, formerly the student of one of the male researchers; a postdoctoral

---

13 I have excluded from my survey any indexed individuals whose gender I could not discern from their name or the text.

student in computer science; and a psychology professor. Only a quarter of these women can be identified as scientists, half are introduced via their connection with male characters, and none are allotted more than a few lines. *The First Three Minutes* indexes over fifty male names and three female; two of these three are scientists, and both are mentioned only once, and then as co-researchers with men. One, Margaret Burbidge, shares the same name as her co-researcher, Geoffrey Burbidge (125) (they are married, although this is not mentioned in the text). *A Brief History of Time* lists close to one hundred male names and one female: Jocelyn Bell, who as a research student discovered neutron stars. Zukav's book also features over one hundred men and one woman; the latter is mentioned once in the text and from the context is presumably a physicist (320). There are also a number of unindexed women: Hawking's wife, Jane Wilde, his daughter Lucy, and his secretary Judy Fella, in *A Brief History of Time*, as well as the "little old lady" with whose comical question the text begins; and several unnamed wives and mothers of scientists in *Chaos* and *Complexity*.

The most obvious explanation for the general non-appearance of women in physics popularizations is the low percentage of women within physics, both at present and historically.[14] This could also explain the fact that there are very few female authors of popular physics books—Christine Sutton, Danah Zohar, Kitty Ferguson and Margaret Wertheim are some exceptions, and of these only Sutton is a working physicist. However, in some cases the exclusion of women is not limited to their non-appearance as scientists. Hayles notes that in Gleick's text none of the men spend time with women or have relationships with women, and (as noted earlier) their living conditions and lifestyles suggest solitariness and anti-domesticity ("Turbulence" 243); thus women's absence is evident even in "the personal realm … where we would expect women to be in the male-dominated world Gleick writes about" (244). Other texts provide similar examples of exclusion: almost all of the popularizers discussed in this book use the male pronoun to describe "the scientist" or "the experimenter," and often use male figures in diagrams.[15] Perhaps these popularizers adopt the same puzzled attitude as Richard Dawkins, with whom it is rather difficult to sympathize when he states that he is "distressed to find that some women friends (fortunately not many) treat the use of the impersonal masculine pronoun as if it showed intention to exclude them" (*The Blind Watchmaker* xvi). Examples of condescension are evident: the woman who is to become Hawking's wife is "a very nice girl" (49), his secretary uses "her not inconsiderable charm" to obtain free airline tickets; Davies in *The Runaway Universe* (1978) describes the

---

14 Haynes quotes a 1992 survey of 403 physics departments throughout the world that found that (in her words) "only 13 percent of faculty members are women" (317 n. 4).

15 There are, however, signs of change. For example, Paul Davies's *About Time*, in contrast to his many previous popularizations, features gender-neutral language: "the experimenter" may change "his or her mind" (169); "she or he" has no control over individual particles (172). In discussing relativity and the twin paradox, Davies previously tended to use astronauts at points "A" and "B" which are referred to as "he" in the text and depicted as male in diagrams. In *About Time* the astronauts become Anne and Betty, and play a substantial part in the text; the typically male pronouns are then effectively reversed: "When you have an infinite timewarp, one woman's microsecond is another woman's forever" (120).

discoverer of the pulsar as "Miss Jocelyn Bell," and slightly later as "Miss Bell," but doesn't give titles to any other researchers (126–7). Some of Feynman's anecdotes, such as the one in which he adopts "the attitude that those bar girls are all bitches, that they aren't *worth* anything" to see if this improves his ability to get them to sleep with him (*Surely You're Joking* 189–90), are, to say the least, objectifying.

While much of this sexism can be attributed to a wider sexism within academia and society in general, it is also possible to speculate on more specific sources. Hayles suggests that Gleick's exclusion of women from science in *Chaos* is designed to combat the powerful femininity associated with the "polysemy" of chaos itself ("Turbulence" 245). However, an equally likely contributing factor in Gleick's case is the adoption of stereotypes that emphasize masculine autonomy, in particular the traditional hard-boiled PI. As Jopi Nyman argues in a gender-based analysis of hard-boiled fiction, the vision of individualism central to this genre is very much a masculinized one: "the genre aims at a reaffirmation of a disrupted masculine social order ... hard-boiled fiction attempts to defend the ideal of the autonomous male: the character is shown to be a truly masculine character who opposes all forms of Otherness ..." (3–4). In these novels, Nyman writes, "the clearly defined and strictly ordered masculine values of the hard-boiled male's world" are "[c]onfronted with the feminine chaos of the modern world" (364). In the hard-boiled genre, women are part of the chaos out of which the PI must make some order—part of the personal and social networks from which he strives to stay aloof. The women who create most chaos in hard-boiled detective narratives are *femmes fatales*: deadly temptresses who seduce and threaten the PI. The adoption of the PI as a stereotype by a popularizer immediately casts women into this kind of complementary role, excluding them from the main role of scientist/detective. It is easy to locate female characters in Waldrop's text who conform readily to the *femme fatale* stereotype. One, an economist, could have stepped out of *The Big Sleep*: "Standing there with a scotch in one hand and a cigarette in the other, she was a most formidable lady ... . She was rumoured to eat bureaucrats for breakfast"; like Chandler's Vivian Reagan, this woman leaves a trail of male lovers in her wake, having "married, in succession, most of the top economists in Hungary" (48). Another woman, the wife of a researcher, is described as "feisty" and "blunt-talking" (216). But when they do not fit this role, women are largely relegated to the background, part of the world of domesticity and personal relationships from which the scientist, like the hard-boiled PI, is isolated.

A similar argument can be applied to the other stereotypes described above. The figure of the scientist-as-priest, for example, is based on a rejection of worldly, material concerns, including the needs and constraints of the physical body. John Michael, in his analysis of Hawking's image, states that "the figure of the scientist in Western popular culture has long been the figure of a disembodied man, or the figure of a man whose attention to the material and social world ... is defined by his distracted attention to the problems of an immaterial or purely intellectual realm." Hawking, he claims, has more than anyone "captured this oddly gendered disembodiment" (207). Michael presumably considers the gendering of disembodiment "odd" because bodies are what usually determine the assignment of gender. And yet in another way the figure of the disembodied man is not odd in the slightest, as feminist critics have long argued that men are not perceived as tied to their bodies in the same way that women are. It is the female body that is marked as gendered, whereas the male body is considered universal, the basic human form:

> A man never begins by presenting himself as an individual of a certain sex; it goes without saying that he is a man ... . Woman has ovaries, a uterus: these peculiarities imprison her in her subjectivity, circumscribe her within the limits of her own nature ... . [Man] thinks of his body as a direct and normal connection with the world, which he believes he apprehends objectively, whereas he regards the body of woman as a hindrance, a prison, weighed down by everything peculiar to it. (de Beauvoir xxi–xxii)

A scientist who is a woman is then always a *woman* scientist, not simply a scientist; and as a woman scientist she cannot be disembodied, as women are defined by their bodies and more specifically by their reproductive capacity ("ovaries, a uterus"). Women have traditionally been associated with motherhood, family, personal relationships, emotionality and domesticity. Thus where a male scientist can be cast in the role of the absent-minded professor, the slightly crazed obsessive or the detached Holmesian detective, for a female scientist these roles would imply a failure to fulfil more deep-seated stereotypes of mother, carer, nurturer, provider of social cohesion.[16] These stereotypes might be applied to a woman scientist, but she would be perceived as masculinized as a result.

This chapter does not, of course, exhaust the various stereotypes of the physicist, or the scientist more generally, which could be examined; it needs to be read alongside examinations of other popularizations (such as Charney's study of *Longitude*) and of scientists' biographies and autobiographies; studies (such as Haynes's) of the representation of the scientist in fiction, film and other media; and also sociological analyses designed to empirically determine which stereotypical features the public tends to absorb.[17] Nonetheless, my analysis does indicate the prevalence in recent prominent popularizations of stereotypes which reinforce the sense that scientists are immune to social constraints and bonds. It is not my claim that popularizers fabricate the characteristics which they attach to their researchers: according to a source independent of Gleick, Feigenbaum *does* consume "countless filterless cigarettes and espressos," just as he *does* resemble Beethoven (Horgan 222); but it is Gleick's decision to include, and to dwell on, these features which constitute his reinforcement of a particular stereotype. The same can be said of the other stereotypes described above. Such stereotypes may be chosen primarily for their appeal to readers, and/or (in Gleick's and Waldrop's cases) their consonance with

---

16 Haynes includes in her book a very telling photograph from the early twentieth century, depicting a scientist in a laboratory peering intently at a flask while a child screams unnoticed at her feet. The photograph, Haynes explains, was taken "to deride Marie Curie" (298).

17 Recent examples include Jrène Rahm and Paul Charbonneau's "Probing Stereotypes through Students' Drawings of Scientists" and Kristina Petkova and Pepka Boyadjieva's "The Image of the Scientist and its Functions."

the kind of science being described. But these stereotypes bring to popularizations more than these qualities: they also construct for the reader a sense of the typical personality and lifestyle of the scientist.

I have focused on this particular set of stereotypes because they relate to a point of contention within the "Science Wars," the relationship between the scientist, science and society. In particular, they relate to debates about the interaction (or lack thereof) between science and the wider culture. An awareness of such stereotypes can help the reader or critic to recognize the covert endorsement of a particular conception of science. More specifically, it can draw attention to groups of people that certain representations may exclude, such as women. It is only through analysis of the discursive means by which such stereotypical representations operate that they can begin to be effectively challenged.

Conclusion

The previous three chapters have dealt with a diverse range of ideas: Dancing Wu Li Masters and decision-making particles; Big Bang mythology; hard-drinking, chain-smoking chaoticians. There is, however, an organizing principle which draws together these apparently disparate analyses: each of the last three chapters highlights the literary aspects of popular physics writing, and thereby identifies the active role that popularizations can play in debates between the literary and scientific communities. In each case, my analysis of the textual strategies deployed in popularizations reveals the wider role these devices can play in cross-disciplinary discussion.

The early chapters of this book establish two relevant forms of cross-disciplinary "skirmish." The first, which applies primarily to Britain, is essentially an extension of Snow's "two cultures": a hostility between literary and scientific intellectuals deriving from long-standing resentments, socially entrenched structures and attitudes, and a lack of shared territory. What is new, I have argued, is the conspicuous role of popular science books in this exchange. The popular science boom of the late twentieth century generated polarized reactions within the literary community. While some novelists and playwrights creatively incorporated ideas derived from popularizations into their work, others vigorously opposed what they saw as popularizers' overweening scientism. The latter have in turn been publicly challenged by a number of prominent popularizers, who believe that the literary intelligentsia tend to monopolize the term "culture" through their control of the media and resent any intrusion by scientists into what they perceive as their domain.

The second form of interdisciplinary disagreement I have outlined is the "Science Wars," heated cross-disciplinary exchanges predominantly (although not wholly) based in the United States. These debates derive primarily from developments within the social sciences—the disciplines which, ironically, Snow viewed as the "third culture" that might act as a mediator between the polarized literary and scientific communities. Unlike the literary intellectuals whose insularity Snow criticized, science studies critics are interested in scientific culture, on an anthropological if not always a technical level—so in this sense, they are well positioned to mediate between the "two cultures." But what Snow could not foresee from his pre-Kuhnian position was the impact of relativist and constructivist attacks on the authority of science. The central issues of the "Science Wars" do not concern the lack of shared knowledge between the "two cultures," but focus on more fundamental epistemological and methodological questions. For constructivist sociologists and philosophers of science, the traditional "two cultures" debate dissolves once one accepts that scientific discourse is culturally embedded in the same way as any other discourse, and thus has no special access to or ability to describe "reality." However, such a view itself generates new cross-disciplinary hostilities.

As the science studies community becomes increasingly perceived by scientists as a challenge to the authority of science (evidenced by the intense public debate sparked by the Sokal Affair), the image of science that the public receives takes on a renewed importance. Thus, although the "Science Wars" are admittedly an esoteric series of debates conducted for the most part within academic communities, one must remember that the issues at stake here are quite concrete and far-reaching. The non-scientific public, and specific non-scientific groups such as government committees, will approach questions concerning science-related policy decisions in quite a different manner if they believe that science is a form of story-telling little different from myth than if they consider it a reliable, objective form of knowledge.

The literary analysis of several key popular physics books presented in the latter part of this study suggests that these popularizations are not neutrally positioned in the skirmishes described above. Popularizations, I have argued, are themselves sites at which cross-disciplinary debate takes place, both overtly in popularizers' explicit comments, and covertly (and perhaps subconsciously) in the textual strategies they employ.

Quantum theory, the focus of my first case study, is an area of physics to which literary critics have been particularly drawn, with post-structuralists citing this field as evidence that the reductionist, objective stance of science has been eroded from within. These critics often rely for their knowledge of scientific concepts on popularizers, in particular New Age popularizers such as Capra and Zukav, whose Eastern mystical perspective leads them at times to oppose the same traditional Western philosophy that post-structuralism challenges. Ironically, while these critics are frequently quick to point to the subjective nature of knowledge, they are sometimes reluctant to take into account the ambiguous language of the popularizations on which they draw. My analysis of the use of metaphor in Zukav's *The Dancing Wu Li Masters* identifies the slippage of meaning that renders this text more of a hindrance than a help to cross-disciplinary communication.

My discussion of popular cosmology shows that popularizers such as Weinberg and Hawking conduct "two cultures" boundary work at the same time as explaining scientific concepts. Both *The First Three Minutes* and *A Brief History of Time* begin by rhetorically dismissing a literary form of knowledge (myth) in favour of science, but paradoxically end by constructing their own brand of teleological myth in which science becomes the ultimate goal of the universe. The mythic narratives of these two physicists can be interpreted not merely as scientific hubris, but also as a defensive reaction against challenges to the boundary between science and myth, challenges which have recently been directed at Big Bang cosmology from within both the humanities and the sciences.

My final analysis, which concentrates on two prominent popularizations of chaos and complexity theories, reveals how popularizers' characterization of scientists feeds into cross-disciplinary debates about the relationship between science and its practitioners. Both Gleick and Waldrop, I argue, adopt the language, atmosphere and character stereotypes of the hard-boiled detective novel. In both cases, the hard-boiled format provides a proven way of attracting reader interest, and also resonates with the kind of science described. However, the PI stereotype (like the

Holmesian detective) also brings with it a sense of social autonomy and isolation that is projected onto the scientist, despite each popularizer's interest in networks rather than individuals. The stereotype of the scientist-as-PI can be grouped with other favourite stock images of scientists that emphasize isolation from cultural conventions, such as the absent-minded professor, the priest and the obsessive. The adoption of these stereotypes opposes the constructivist view of scientists as embedded in and to a large extent determined by the culture in which they operate. Unsurprisingly, when commentators such as Lewis Wolpert and Freeman Dyson voice this opposition explicitly, it is on roughly the same terms. Scientists are "[i]nnocents and outsiders," people who are immune to the "local tyranny" imposed by culture (Wolpert and Richards 2; Dyson 2). My analysis shows that literary criticism is well-positioned to analyse the various images of the scientist constructed by popularizers, and to recognize their relevance to cross-disciplinary debate.

I have demonstrated, then, how the appropriation of scientific concepts by literary critics can reveal the unstable nature of metaphor as an expository tool; how the narrative structure of popular physics books can encode a debate about the nature of literary and scientific knowledge; and how the popularizers' use of character types borrowed from fiction reflects a characterization of science itself. I hope, in each case, that my observations are interesting in themselves; but even more, I hope they make the broader point that, far from being a mere distortion of professional science, or a threatening new presence on the bestseller lists, these popular physics books represent very fertile ground for literary critics interested in the interaction of scientific and non-scientific cultures—whether this is framed as a hostile or a harmonious exchange.

Reading Popular Physics represents an excursion into a relatively empty field. It is not a systematic analysis of the conventions of popular physics books, but an overview of the genre's development and cultural significance, and an examination of a number of prominent examples of this genre within a specific context (the "Science Wars" and related cross-disciplinary debates). My approach is intended to encourage, rather than pre-empt, more research into this culturally significant genre. There are numerous avenues open to investigation that I have hardly touched upon here. For example, I have looked at the novelistic aspects of popular physics books but avoided in-depth discussion of the inverse phenomenon: the increasingly frequent appearance of popular physics in fiction. I have also explicitly avoided the question of the interaction between science fiction and popular science, even though some of the most famous popularizers, such as Isaac Asimov and Arthur C. Clarke, are equally famous as science fiction writers, and a number of the popularizers I have discussed, such as Paul Davies and John Gribbin, have also tried their hand at the genre. I have avoided this subject not because it is insignificant, but because it is so significant that it deserves far more than the token-gesture page or two I could afford.[1] The same argument applies to the relationship between popular and

---

1 For analyses that deal with the science fiction/science popularization interface, see my article "Reading Aldiss and Penrose's *White Mars* as Science Faction" and Mellor's "Between Fact and Fiction."

professional science. The question of how their relationship is to be formulated so as to avoid a reinscription of the traditional bipolar model, and how feedback processes between them can be understood, requires detailed investigation of what is lost, gained and rearranged in the translation of specialized scientific language and mathematics into the metaphor-laden discourse of popularization. And, most obviously, popular books are one of many different forms of popularization, and popular physics is only one of numerous subgenres that make up popular science; all of these facets of science popularization are ripe for analysis.

I will have achieved my aim if my explorations provide the impetus for other investigations into these areas. It is an academic cliché to humbly offer one's research as primarily "a springboard for further analysis," but in this case there is a strong imperative behind this cliché: for until "literature and science" criticism develops a substantial body of research into the discourses of science popularization, it cannot claim to have a firm grip on the so-called "two cultures" question.

Bibliography

Popularizations

Asimov, Isaac. *In the Beginning*. London: New English Library, 1981.
——. *Inside the Atom*. London: Abelard-Schuman, 1956.
Atkins, Peter. *The Creation*. Oxford: Freeman, 1981.
Barrow, John. *Theories of Everything: The Quest for Ultimate Explanation*. London: Vintage, 1992.
Bell, J. S. *Speakable and Unspeakable in Quantum Mechanics*. Cambridge: Cambridge University Press, 1987.
Bohm, David. *Wholeness and the Implicate Order*. London: Routledge, 1982.
Bohr, Niels. *Atomic Physics and Human Knowledge*. New York: Wiley; London: Chapman, 1958.
——. *Atomic Theory and the Description of Nature*. Cambridge: Cambridge University Press, 1934.
——. *Essays 1958–1962 on Atomic Physics and Human Knowledge*. New York: Interscience-Wiley, 1963.
Born, Max. *The Restless Universe*. Trans. Winifred M. Deans. London: Blackie, 1935.
Boslough, John. *Masters of Time: How Wormholes, Snakewood and Assaults on the Big Bang Have Brought Mystery Back to the Cosmos*. London: Phoenix, 1993.
Braunbek, Werner. *The Drama of the Atom*. Trans. Brian J. Kenworthy and W. A. Coupe. Edinburgh: Oliver, 1958.
Briggs, John P. and F. David Peat. *Looking Glass Universe: The Emerging Science of Wholeness*. London: Fontana, 1985.
Brockman, John. *The Third Culture: Beyond the Scientific Revolution*. New York: Simon, 1995.
Bronowski, Jacob. *Ascent of Man*. London: BBC, 1973.
Bruce, Colin. *The Strange Case of Mrs Hudson's Cat Or Sherlock Holmes Solves the Einstein Mysteries*. London: Vintage, 1998.
Bryson, Bill. *A Short History of Nearly Everything*. London: Doubleday, 2003.
Calder, Nigel. *The Key to the Universe*. London: BBC, 1977.
Capra, Fritjof. *The Tao of Physics: An Exploration of the Parallels Between Modern Physics and Eastern Mysticism*. 1975. London: Flamingo-HarperCollins, 1992.
Carson, Rachel. *The Sea Around Us*. 1951. London: Panther, 1965.
——. *Silent Spring*. Greenwich, CT: Fawcett, 1962.
Casti, John. *Complexification: Explaining a Paradoxical World Through the Science of Surprise*. London: Abacus, 1994.
Chown, Marcus. *Afterglow of Creation: From the Fireball to the Discovery of Cosmic Ripples*. London: Arrow, 1993.
Cohen, Jack, and Ian Stewart. *The Collapse of Chaos: Discovering Simplicity in a Complex World*. London: Viking, 1994.

Compton, A. *Atomic Quest: A Personal Narrative*. London: Oxford University Press, 1956.
Coveney, Peter, and Roger Highfield. *Frontiers of Complexity: The Search for Order in a Chaotic World*. London: Faber, 1996.
Davies, Paul. *About Time: Einstein's Unfinished Revolution*. London: Viking, 1995.
———. *The Cosmic Blueprint: Order and Complexity at the Edge of Chaos*. 1987. London: Penguin, 1995.
———. *The Edge of Infinity: Beyond the Black Hole*. London: Penguin, 1994.
———. *The Forces of Nature*. 1979. 2nd edn. Cambridge: Cambridge University Press, 1986.
———. *God and the New Physics*. 1983. London: Penguin, 1986.
———. *The Last Three Minutes: Conjectures About the Ultimate Fate of the Universe*. London: Phoenix, 1994.
———. *The Mind of God: Science and the Search for Ultimate Meaning*. London: Penguin, 1993.
———. "The New Physics: A Synthesis." *The New Physics*. Ed. Paul Davies. Cambridge: Cambridge University Press, 1989. 1–6.
———. *Other Worlds: Space, Superspace and the Quantum Universe*. London: Dent, 1980.
———. *Runaway Universe*. London: Dent, 1978.
———. *Superforce: The Search for a Grand Unified Theory of Nature*. 1984. London: Penguin, 1995.
——— and Julian Brown. *Superstrings: A Theory of Everything*. Cambridge: Canto-Cambridge University Press, 1992.
——— and John Gribbin. *The Matter Myth: Beyond Chaos and Complexity*. 1991. London: Penguin, 1992.
Dawkins, Richard. *The Blind Watchmaker*. London: Penguin, 1988.
———. *River Out of Eden: A Darwinian View of Life*. London: Phoenix, 1996.
———. *The Selfish Gene*. New York: Oxford University Press, 1978.
———. *Unweaving the Rainbow: Science, Delusion and the Appetite for Wonder*. London: Allen Lane-Penguin, 1998.
de Broglie, Louis. *The Revolution in Physics: A Non-Mathematical Survey of Quanta*. Trans. Ralph W. Niemeyer. London: Routledge, 1954.
Dietz, David. *Atomic Energy Now and Tomorrow*. London: Westhouse, 1946.
Dingle, Herbert. *Relativity for All*. London: Methuen, 1922.
Disney, Michael. *The Hidden Universe*. London: Dent, 1984.
Eddington, A. S. *The Nature of the Physical World*. 1928. Cambridge: Cambridge University Press, 1931.
———. *The Philosophy of Physical Science*. 1939. Cambridge: Cambridge University Press, 1949.
———. *Science and the Unseen World*. London: Allen, 1929.
———. *Space, Time and Gravitation: An Outline of General Relativity*. 1920. Cambridge, Cambridge University Press, 1987.
Eidinoff, Maxwell Leigh and Hyman Ruchlis. *Atomics*. London: Harrap, 1950.
Einstein, Albert. *Relativity: The Special and General Theory*. Trans. Robert W. Lawson. London: Methuen, 1920.

——— and Leopold Infeld. *The Evolution of Physics: The Growth of Ideas from Early Concepts to Relativity and Quanta*. New York: Simon, 1938.

Faraday, Michael. *A Course of Six Lectures on the Chemical History of a Candle*. 1861. Ed. William Crookes. London: Scientific Book Guild-Beaverbrook Newspapers, 1960.

Feinberg, J. G. *The Atom Story: Being the Story of the Atom and the Human Race*. London: Wingate, 1952.

Ferguson, Kitty. *The Fire in the Equation: Science, Religion, and the Search for God*. London: Bantam, 1994.

Feynman, Richard P. *The Character of Physical Law*. 1965. London: Penguin, 1992.

———. *The Meaning of It All*. London: Penguin, 1998.

———. *QED: The Strange Theory of Light and Matter*. 1985. London: Penguin, 1990.

———. *"Surely You're Joking, Mr Feynman!" Adventures of a Curious Character*. 1985. Ed. Edward Hutchings. London: Vintage, 1992.

———. *"What Do You Care What Other People Think?" Further Adventures of a Curious Character*. 1988. London: Grafton-HarperCollins, 1992.

Frisch, Otto R. *Atomic Physics Today*. 1961. Edinburgh: Oliver and Boyd, 1962.

———. *Meet the Atoms: A Popular Guide to Modern Physics*. London: Sigma, 1947.

Gamow, George. *Atomic Energy in Cosmic and Human Life: Fifty Years of Radioactivity*. Cambridge: Cambridge University Press, 1947.

———. *The Creation of the Universe*. 1952. Rev. edn. London: Macmillan, 1961.

———. *Mr Tompkins Explores the Atom*. 1944. Cambridge: Cambridge University Press, 1945.

———. *Mr Tompkins in Paperback*. 1965. Cambridge: Cambridge University Press, 1990.

———. *Mr Tompkins in Wonderland or Stories of c, G and h*. Cambridge: Cambridge University Press, 1939.

Gardner, Martin. *The Ambidextrous Universe*. 1964. London: Allen Lane-Penguin, 1967.

———. *Relativity for the Million*. New York: Macmillan, 1962.

Gell-Mann, Murray. *The Quark and the Jaguar: Adventures in the Simple and the Complex*. London: Little, 1994.

Gilmore, Robert. *Alice in Quantumland: An Allegory of Quantum Physics*. Wilmslow: Sigma Science, 1994.

Gleick, James. *Chaos: Making a New Science*. 1987. London: Penguin, 1988.

———. *Genius: Richard Feynman and Modern Physics*. London: Little, 1996.

Goswami, Amit, with Richard E. Reed and Maggie Goswami. *The Self-Aware Universe: How Consciousness Creates the Material World*. London: Simon, 1993.

Gould, Stephen Jay. *The Mismeasure of Man*. 1981. London: Penguin, 1992.

Greene, Brian. *The Elegant Universe: Superstrings, Hidden Dimensions, and the Quest for the Ultimate Theory*. 1999. London: Vintage-Random, 2000.

Gribbin, John. *Blinded by the Light: The Secret Life of the Sun*. London: Bantam, 1991.

———. *In Search of the Big Bang: Quantum Physics and Cosmology*. London: Corgi, 1988.

———. *In Search of Schrödinger's Cat: Quantum Physics and Reality*. 1984. London: Black Swan, 1991.

———. *In the Beginning: The Birth of the Living Universe*. London: Viking, 1993.

Guth, Alan. *The Inflationary Universe: The Quest for a New Theory of Cosmic Origins.* London: Cape, 1997.

Hawking, Stephen W. *Black Holes and Baby Universes and Other Essays.* London: Bantam, 1994.

——. *A Brief History of Time: From the Big Bang to Black Holes.* 1988. London: Bantam, 1989.

——, ed. *Stephen Hawking's* A Brief History of Time*: A Reader's Companion.* Prepared by Gene Stone. London: Bantam, 1992.

—— with Leonard Mlodinow. *A Briefer History of Time.* London: Bantam, 2005.

Hecht, Selig. *Explaining the Atom.* 1947. Rev. edn. with additional material by Eugene Rabinowitch. London: Gollancz, 1955.

Heisenberg, Werner. *Physics and Philosophy: The Revolution in Modern Science.* London: Allen, 1959.

Herbert, Nick. *Elemental Mind: Human Consciousness and the New Physics.* New York: Dutton, 1993.

——. *Quantum Reality: Beyond the New Physics.* London: Rider, 1985.

Hoffmann, Banesh. *The Strange Story of the Quantum: An Account for the General Reader of the Growth of the Ideas Underlying Our Present Atomic Knowledge.* 1947. 2nd edn. New York: Dover, [*c.* 1980].

Horgan, John. *The End of Science: Facing the Limits of Knowledge in the Twilight of the Scientific Age.* Reading, MA: Helix-Addison-Wesley, 1996.

Huxley, Thomas H. *Collected Essays.* 9 vols. London: Macmillan, 1893–1894.

Jeans, James. *Atomicity and Quanta.* Cambridge: Cambridge University Press, 1926.

——. *The Mysterious Universe.* Cambridge: Cambridge University Press, 1930.

——. *Physics and Philosophy.* Cambridge: Cambridge University Press, 1942.

——. *The Universe Around Us.* Cambridge: Cambridge University Press, 1929.

Johnson, George. *Fire in the Mind: Science, Faith and the Search for Order.* London: Viking, 1996.

Josephson, B. D. and V. S. Ramachandran. *Consciousness and the Physical World.* Oxford: Pergamon, 1980.

Kaku, Michio and Jennifer Thompson. *Beyond Einstein: The Cosmic Quest for the Theory of the Universe.* 1987. Oxford: Oxford University Press, 1997.

Kellert, Stephen. *In the Wake of Chaos: Unpredictable Order in Dynamic Systems.* Chicago: University of Chicago Press, 1993.

Krauss, Lawrence. *The Fifth Essence: The Search for Dark Matter in the Universe.* London: Hutchinson, 1989.

Lederman, Leon with Dick Teresi. *The God Particle.* London: Bantam, 1993.

Lerner, Eric. *The Big Bang Never Happened: A Startling Refutation of the Dominant Theory of the Origin of the Universe.* London: Simon, 1992.

Lewin, R. *Complexity: Life at the Edge of Chaos.* London: Dent, 1993.

Lindley, David. *The End of Physics: The Myth of a Unified Theory.* New York: Basic, 1993.

Lodge, Oliver. *Atoms and Rays: An Introduction to Modern Views on Atomic Structure and Radiation.* London: Benn, 1924.

——. *Ether and Reality: A Series of Discourses on the Many Functions of the Ether of Space.* London: Hodder, 1925.

Lorenz, Edward. *The Essence of Chaos*. London: UCL, 1995.

Mandelbrot, Benoit. *Fractals: Form, Chance, and Dimension*. New York: Freeman, 1977.

———. *The Fractal Geometry of Nature*. New York: Freeman, 1983.

Mather, John C. and John Boslough. *The Very First Light: The True Inside Story of the Scientific Journey Back to the Dawn of the Universe*. London: Penguin, 1998.

McEvoy, J. P. and Oscar Zarate. *Stephen Hawking for Beginners*. Cambridge: Icon, 1995.

Meitner, Lise. Introduction. Frisch, *Meet the Atoms* v–vi.

Morris, Richard. *The Edges of Science: Crossing the Boundary from Physics to Metaphysics*. London: Fourth Estate, 1992.

Overbye, Dennis. *Lonely Hearts of the Cosmos: The Story of the Scientific Quest for the Secret of the Universe*. London: Macmillan, 1991.

Pagels, Heinz R. *The Cosmic Code: Quantum Physics as the Language of Nature*. 1982. London: Penguin, 1994.

Parker, Barry. *Invisible Matter and the Fate of the Universe*. New York: Plenum, 1989.

———. *Search for a Supertheory: From Atoms to Superstrings*. New York: Plenum, 1987.

Peat, F. David. *Blackfoot Physics: A Journey into the Native American Universe*. London: First Estate, 1995.

———. *Superstrings and the Search for the Theory of Everything*. London: Scribner's, 1991.

Penrose, Roger. *The Emperor's New Mind: Concerning Computers, Minds, and the Laws of Physics*. London: Vintage, 1990.

———. *Shadows of the Mind: A Search for the Missing Science of Consciousness*. Oxford: Oxford University Press, 1994.

Polkinghorne, John. *Beyond Science: The Wider Human Context*. Cambridge: Cambridge University Press, 1996.

———. *One World: The Interaction of Science and Theology*. London: SPCK, 1986.

———. *The Quantum World*. London: Penguin, 1990.

———. *The Way the World Is: The Christian Perspective of a Scientist*. London: Triangle, 1983.

Preston, Richard. *The Hot Zone*. London: Corgi, 1995.

Prigogine, Ilya, and Isabelle Stengers. *Order Out of Chaos: Man's New Dialogue with Nature*. 1984. London: Flamingo-Fontana, 1985.

Rae, Alistair I. M. *Quantum Physics: Illusion or Reality*. Cambridge: Cambridge University Press, 1986.

Riordan, Michael. *The Hunting of the Quark: A True Story of Modern Physics*. New York: Touchstone-Simon, 1987.

——— and David N. Schramm. *The Shadows of Creation: Dark Matter and the Structure of the Universe*. Oxford: Oxford University Press, 1993.

Romer, Alfred. *The Restless Atom*. 1960. London: Heinemann, 1961.

Rowan-Robinson, Michael. *Ripples in the Cosmos: A View Behind the Scenes of the New Cosmology*. Oxford: Spektrum-Freeman, 1993.

Ruelle, David. *Chance and Chaos*. London: Penguin, 1993.

Russell, Bertrand. *ABC of Atoms*. London: Routledge, 1923.

———. *ABC of Relativity*. 1925. London: Kegan Paul; New York: Harper, 1926.

Sagan, Carl. *Cosmos*. 1980. London: Macdonald, 1983.

Schrödinger, Erwin. *Mind and Matter. What is Life? The Physical Aspect of the Living Cell & Mind and Matter.* 1958. Cambridge: Cambridge University Press, 1967.
Silk, Joseph. *The Big Bang.* 1980. Rev. edn. New York: Freeman, 1989.
Singh, Simon. *Fermat's Last Theorem: The Story of a Riddle that Confounded the World's Greatest Minds for 358 Years.* London: Fourth Estate, 1997.
Smoot, George, and Keay Davidson. *Wrinkles in Time.* London: Little, 1993.
Sobel, Dava. *Longitude: The True Story of a Lone Genius Who Solved the Greatest Scientific Problem of His Time.* London: Fourth Estate, 1995.
Squires, Euan. *Conscious Mind in the Physical World.* Bristol: Hilger, 1990.
———. *The Mystery of the Quantum World.* 1986. 2nd edn. Bristol: Institute of Physics, 1994.
Stannard, Russell. *Uncle Albert and the Quantum Quest.* London: Faber, 1994.
Stapp, Henry P. *Mind, Matter, and Quantum Mechanics.* Berlin: Springer, 1993.
Stenger, Victor. *The Unconscious Quantum: Metaphysics in Modern Physics and Cosmology.* Amherst, NY: Prometheus, 1995.
[Stewart, B. and P. G. Tait.] *The Unseen Universe or Physical Speculations on a Future State.* 4th edn. London: Macmillan, 1875.
Stewart, Ian. *The Magical Maze: Seeing the World Through Mathematical Eyes.* London: Weidenfeld, 1997.
Sullivan, J. W. N. *Aspects of Science.* London: R. Cobden-Sanderson, [1923].
———. *Atoms and Electrons.* London: Hodder, 1923.
———. *Limitations of Science.* 1933. Clifton, NJ: A. M. Kelley, 1973.
———. *Three Men Discuss Relativity.* London: Collins, 1925.
Talbot, Michael. *Beyond the Quantum.* New York: Macmillan, 1986.
———. *Mysticism and the New Physics.* London: Routledge, 1981.
Taylor, John. *Black Holes: The End of the Universe?* 1973. Glasgow: Fontana-Collins, 1975.
Tilby, Angela. *Science and the Soul: New Cosmology, the Self and God.* London: SPCK, 1992.
Tipler, Frank J. *The Physics of Immortality: Modern Cosmology, God and the Resurrection of the Dead.* 1994. London: Macmillan, 1995.
Trefil, James S. *The Dark Side of the Universe: A Scientist Explores the Mysteries of the Cosmos.* New York: Scribner's, 1988.
Tyndall, John. *Fragments of Science for Unscientific People: A Series of Detached Essays, Lectures and Reviews.* London: Longmans, 1871.
———. *Heat: A Mode of Motion.* 1863. 6th edn. London: Longmans, 1908.
Waldrop, M. Mitchell. *Complexity: The Emerging Science at the Edge of Order and Chaos.* New York: Simon, 1992.
Weinberg, Steven. *Dreams of a Final Theory: The Search for the Fundamental Laws of Nature.* London: Vintage, 1993.
———. *The First Three Minutes: A Modern View of the Origin of the Universe.* 1977. London: Fontana-Collins, 1983.
Whitehead, Alfred North. *Science and the Modern World.* 1925. London: Free Association, 1985.
Wolf, Fred Alan. *Mind and the New Physics.* London: Heinemann, 1985.

———. *Taking the Quantum Leap: The New Physics for Nonscientists*. New York: Harper, 1981.
Zohar, Danah. *The Quantum Self.* 1990. London: Flamingo-HarperCollins, 1991.
——— and Ian Marshall. *The Quantum Society: Mind, Physics and a New Social Vision*. London: Bloomsbury, 1993.
Zukav, Gary. *The Dancing Wu Li Masters: An Overview of the New Physics*. 1979. London: Rider, 1991.

Critical and Other Works

A Brief History of Time. Dir. Errol Morris. Triton Pictures, 1991.
Ackrill, Kate, ed. *The Role of the Media in Science Communication*. Ciba Foundation Discussion Meeting, Dec. 1993. London: Ciba Foundation, [c. 1994].
Agassi, Joseph. "The Detective Novel and Scientific Method." *Poetics Today* 3.1 (1982): 99–108.
Alfvén, H. "Cosmology—Myth or Science?" *Journal of Astrophysics and Astronomy* 5 (1984): 79–98. Rpt. in *IEEE Transactions on Plasma Science* 20 (1992): 590–600.
"All Good Things." *Star Trek: The Next Generation*. Series Seven. Dir. Winrich Kolbe. Videocassette. Paramount Pictures, 1994.
Amis, Martin. *London Fields*. London: Penguin, 1990.
———. *Night Train*. London: Cape, 1997.
Anderson, Alun, and Tim Lincoln. "Million-Dollar Quark." *Nature* 348 (1990): 102.
Anderson, Wilda C. *Between the Library and the Laboratory: The Language of Chemistry in Eighteenth-Century France*. Baltimore: Johns Hopkins University Press, 1984.
Angyal, Andrew. "Loren Eiseley's *Immense Journey*: The Making of a Literary Naturalist." McRae, *Literature of Science* 54–72.
Appleyard, Brian. "God and the Scientists Join Hands in a Quantum Leap." *Sunday Times* 3 June 1990, sec. 3 ("News Review"): 7.
———. "Mighty Minds that Bridge the Worlds of Science and Literature." *Sunday Times* 11 Apr. 1999, sec. 5 ("News Review"): 4.
———. *Understanding the Present: Science and the Soul of Modern Man*. London: Pan, 1993.
Arden, Rosalind. *God Only Knows: Science, Religion and the Search for the Meaning of Everything*. London: Channel 4 Television, 1995.
Asimov, Isaac. "Popularizing Science." *Nature* 306 (1983): 119.
Astore, William J. *Observing God: Thomas Dick, Evangelicalism, and Popular Science in Victorian Britain and America*. Aldershot: Ashgate, 2001.
Atwood, Margaret. *Cat's Eye*. London: Virago, 1990.
———. *Oryx and Crake*. 2003. London: Virago, 2004.
Axon, T. J. *Beyond the Tao of Physics: Mysticism and Modern Physics——A Reappraisal*. Poynton, Stockport, Cheshire: Tehuti, 1988.
Baldick, Chris. *The Concise Oxford Dictionary of Literary Terms*. Oxford: Oxford University Press, 1991.

Baringer, Philip S. "Introduction: The 'Science Wars.'" *After the Science Wars*. Ed. Keith M. Ashman and Philip S. Baringer. London: Routledge, 2001. 1–13.

Barrow, John. "The Analogy of Nature." *Mission to Abisko: Stories and Myths in the Creation of Scientific Truth.'* Ed. John L. Casti and Anders Karlqvist. Reading, MA: Perseus, 1999. 1–6.

Barthes, Roland. *Mythologies*. Trans. Annette Lavers. London: Paladin, 1973.

Basalla, George. "Pop Science: The Depiction of Science in Popular Culture." *Science and its Public: The Changing Relationship*. Ed. Gerald Holton and William A. Blanpied. Boston Studies in the Philosophy of Science 33. Dordrecht: Reidel, 1976. 260–78.

Bascon, William. "The Forms of Folklore: Prose Narratives." Dundes 5–29.

Bazerman, Charles. *Shaping Written Knowledge: The Genre and Activity of the Experimental Article in Science*. Madison: University of Wisconsin Press, 1988.

Beer, Gillian. *Darwin's Plots: Evolutionary Narrative in Darwin, George Eliot and Nineteenth-Century Fiction*. 1983. 2nd edn. Cambridge: Cambridge University Press, 2000.

———. "Eddington and the Idiom of Modernism." *Science, Reason and Rhetoric*. Ed. Henry Krips, J. E. McGuire, and Trevor Melia. Pittsburgh: University of Pittsburgh Press, 1995. 295–315.

———. *Open Fields: Science in Cultural Encounter*. Oxford: Clarendon, 1996.

Beller, Mara. "The Sokal Hoax: At Whom Are We Laughing?" *Physics Today* 51.9 (1998): 29–34.

Bensaude-Vincent, Bernadette. "A Genealogy of the Increasing Gap Between Science and the Public." *Public Understanding of Science* 10 (2001): 99–113.

———. "In the Name of Science." *Science in the Twentieth Century*. Ed. John Krige and Dominique Pestre. Amsterdam: Harwood, 1997. 319–38.

———, and Anne Rasmussen. *La science populaire dans la presse et l'édition: XIX^{eet} XXe siècles*. Paris: CNRS editions, 1997.

Bernstein, Jeremy. "Popular Science." Rev. of *The Dancing Wu Li Masters*, by Gary Zukav, *The Ambidextrous Universe*, by Martin Gardner, and *Mathematics Today*, ed. Lynn Arthur Steen. *New Yorker* 8 Oct. 1979: 169–77.

———. *Quantum Profiles*. Princeton: Princeton University Press, 1991.

Best, Steven. "Chaos and Entropy: Metaphors in Postmodern Science and Social Theory." *Science as Culture* 2.11 (1991): 188–226.

"Bestsellers from Oxford." *New Scientist* 20 Jan. 1996: 44.

Binyon, T. J. *'Murder Will Out': The Detective in Fiction*. Oxford: Oxford University Press, 1989.

The Black Hole. Dir. Gary Nelson. Walt Disney, 1979.

Black, Joel. "Introduction." Lee and Slade 131–9.

Black, Max. "More About Metaphor." Ortony, *Metaphor and Thought* 19–41.

Blackett, P. M. S. "Memories of Rutherford." *Rutherford at Manchester*. Ed. J. B. Birks. London: Heywood, 1962. 102–113.

Blinderman, Charles S. "Semantic Aspects of T. H. Huxley's Literary Style." *Journal of Communication* 12.3 (1962): 171–8.

Bohnenkamp, Dennis. "Post-Einsteinian Physics and Literature: Toward a New Poetics." *Mosaic* 22.3 (1989): 19–30.

Booker, M. Keith. "Joyce, Planck, Einstein, and Heisenberg: A Relativistic Quantum Mechanical Discussion of *Ulysses*." *James Joyce Quarterly* 27.3 (1990): 577–86.

Boyd, Richard. "Confirmation, Semantics, and the Interpretation of Scientific Theories: Introductory Essay." Boyd, Gasper and Trout 3–35.

———. "Metaphor and Theory Change: What is 'Metaphor' a Metaphor For?" Ortony, *Metaphor and Thought* 481–532.

———. "On the Current Status of Scientific Realism." Boyd, Gasper and Trout 195–222.

———, Philip Gasper, and J. D. Trout, eds. *The Philosophy of Science*. Cambridge, MA: MIT Press, 1991.

Bradshaw, David. "The Best of Companions: J. W. N. Sullivan, Aldous Huxley, and the New Physics." *Review of English Studies* 47 (1996): 188–206; 352–68.

Bragg, Melvyn. "Whose Side Are You On?" *Observer* 7 Mar. 1999, "Review": 1–2.

Briggs, Asa. "The 1990s: The Final Chapter." *Fins de Siècle: How Centuries End, 1400–2000*. Ed. Asa Briggs and Daniel Snowman. New Haven: Yale University Press, 1996. 197–233.

Brissenden, Alan. *Shakespeare and the Dance*. London: Macmillan, 1981.

Brockman, John. "Science Writers: Facts and Fiction." *Times Higher Education Supplement* 31 May 1996: 13.

Broderick, Damien. *The Architecture of Babel: Discourses of Literature and Science*. Melbourne: Melbourne University Press, 1994.

Brooke, John Hedley. *Science and Religion: Some Historical Perspectives*. Cambridge: Cambridge University Press, 1991.

Brooks, G. P. "Mental Improvement and Vital Piety: Isaac Watts and the Benefits of Astronomical Study." *Dalhousie Review* 65 (1985–86): 551–64.

Bruce, Donald, and Anthony Purdy, eds. *Literature and Science*. Rodopi Perspectives on Modern Literature 14. Amsterdam: Rodopi, 1994.

Brush, Stephen G. "Alfvén's Programme in Solar System Physics." *IEEE Transactions on Plasma Science* 20 (1992): 577–89.

———. *The Temperature of History: Phases of Science and Culture in the Nineteenth Century*. New York: Franklin, 1978.

Bryson, Michael A. "Nature, Narrative, and the Scientist-Writer: Rachel Carson's and Loren Eiseley's Critique of Science." *Technical Communication Quarterly* 12.4 (2003): 369–87.

Bucchi, Massimiano. *Science and the Media: Alternative Routes in Scientific Communication*. Routledge Studies in Science, Technology and Society 1. London: Routledge, 1998.

Buchanan, Rex. "Books and the Popularization of Science." *Publishing Research Quarterly* 7 (1991): 5–10.

Burnham, John C. *How Superstition Won and Science Lost: Popularizing Science and Health in the United States*. New Brunswick: Rutgers University Press, 1987.

Bushnell, Jack. "A Brief History of Stephen Hawking: Beginnings and Endings." *Michigan Quarterly Review* 39.4 (2000): 678–90.

Butler, Stephen. Telephone interview. 8 Dec. 1998.

Byatt, A. S. "Belief in the Jungle of Ideas." Rev. of *Consilience*, by Edward O. Wilson. *Guardian* 29 Aug. 1998, "Saturday Review": 8.

Byrne, Kieran R. "The Royal Dublin Society and the Advancement of Popular Science in Ireland, 1731–1860." *History of Education* 15.2 (1986): 81–8.

Calsamiglia, Helena ed. *Popularization Discourse*. Spec. issue of *Discourse Studies* 5.2 (2003): 139–279.

Carey, John ed. *The Faber Book of Science*. London: Faber, 1995.

——. "A Tale of Two Cultures." *Prospect* Nov. 1995: 38–43.

Carson, Cathryn. "Who Wants a Postmodern Physics?" *Science in Context* 8.4 (1995): 635–55.

Cassidy, Angela. "Popular Evolutionary Psychology in the UK: An Unusual Case of Science in the Media?" *Public Understanding of Science* 14 (2005): 115–41.

Cathode, Ray [pseud.]. "Pop Science." *Sight and Sound* 8.3 (1998): 32.

Chandler, Raymond. *The Big Sleep*. 1939. *Raymond Chandler: Three Novels*. London: Penguin, 1993. 1–164.

Charlesworth, Kate. "Life the Universe & (almost) Everything: Novel Values." Cartoon. *New Scientist* 11 Jan. 1992: 55.

Charney, Davida. "Lone Geniuses in Popular Science: The Devaluation of Scientific Consensus." *Written Communication* 20.3 (July 2003): 215–41.

——, ed. *The Rhetoric of Popular Science*. Spec. issue of *Written Communication* 21.1 (2004): 3–105.

Chopra, Deepak. *Quantum Healing: Exploring the Frontiers of Mind/Body Medicine*. New York: Bantam, 1989.

Christie, John and Sally Shuttleworth, eds. and introd. *Nature Transfigured: Science and Literature, 1700–1900*. Manchester: Manchester University Press, 1989.

Cloître, Michel and Terry Shinn. "Expository Practice: Social, Cognitive and Epistemological Linkage." Shinn and Whitley 31–60.

Cole, Stephen. *Making Science: Between Nature and Society*. Cambridge, MA: Harvard University Press, 1992.

Collini, Stefan. Introduction. Snow vii–lxxi.

Collins, Harry and Trevor Pinch. *The Golem: What Everyone Should Know About Science*. Cambridge: Cambridge University Press, 1995.

Comte, Auguste. *The Positive Philosophy of Auguste Comte*. Trans. Harriet Martineau. 2 vols. London: John Chapman, 1853.

Connor, Steve. "Crossing the Great Divide." *Sunday Times* 8 Sept. 1996, sec. 10 ("Culture"): 6–7.

Cooter, Roger. *The Cultural Meaning of Popular Science: Phrenology and the Organization of Consent in Nineteenth-Century Britain*. Cambridge: Cambridge University Press, 1984.

—— and Stephen Pumfrey. "Separate Spheres and Public Places: Reflections on the History of Science Popularization and Science in Popular Culture." *History of Science* 32 (1994): 237–67.

COPUS (Committee on the Public Understanding of Science). "The Book, the Niche, the Strife and the Cover: Popular Science Writing in the 1990s." Leaflet accompanying a session of the British Association Annual Festival of Science, 11 Sept. 1996.

Cordle, Daniel. *Postmodern Postures: Literature, Science and the Two Cultures Debate*. Aldershot: Ashgate, 1999.

Coupe, Laurence. *Myth*. London: Routledge, 1997.
Crocker, D. R. "Anthropomorphism: Bad Practice, Honest Prejudice?" *New Scientist* 16 July 1981: 159–62.
Crocker, Joe. "Give the Public a Break!" *New Scientist* 11 Dec. 1996: 56–7.
Crosland, Maurice. "Popular Science and the Arts: Challenges to Cultural Authority in France under the Second Empire." *British Journal for the History of Science* 34 (2001): 301–322.
Crowden, Alan. "In Pursuit of the Hawking." *New Scientist* 4 June 1994: 41.
Curtis, Ron. "Narrative Form and Normative Force—Baconian Storytelling in Popular Science." *Social Studies of Science* 24 (1994): 419–61.
Daum, Andreas. *Wissenschaftpopularisierung im 19. Jahrhundert: Bürgerliche Kultur, Naturwissenschaftliche Bildung und die Deutsche Öffentlichkeit*, 1848–1914. Munich: R. Oldernbourg, 1998.
Davies, Paul. "The Arts Have Lost It." *Sunday Times* 18 Aug. 1996, sec. 10 ("Culture"): 12–13.
———. Email to the author. 16 March 2006.
———. Personal interview. 19 January 1996.
Dawkins, Richard. "Postmodernism Disrobed." Rev. of *Intellectual Impostures*, by Alan Sokal and Jean Bricmont. *Nature* 394 (1998): 141–3.
Dear, Peter, ed. *The Literary Structure of Scientific Argument*. Philadelphia: University of Philadelphia Press, 1991.
de Beauvoir, Simone. *The Second Sex*. 1953. Trans. H. M. Parshley. New York: Vintage-Random, 1989.
de Camp, L. Sprague and Thomas Clareson. "The Scientist." *Science Fiction: A Contemporary Mythology*. Ed. Patricia S. Warrick, Martin H. Greenberg, and Joseph D. Olander. New York: Harper and Row, 1978. 196–206.
Delaney, Paul. *Tom Stoppard: The Moral Vision of the Major Plays*. Basingstoke: Macmillan, 1990.
DePorter, Bobbi with Mike Hernacki. *Quantum Learning: Unleash the Genius Within You*. London: Piatkus, 1993.
"Descent." *Star Trek: The Next Generation*. Series Six. Dir. Alexander Singer. Videocassette. Paramount Pictures, 1993.
Dickson, David. "The 'Sokal Affair' Takes Transatlantic Turn." *Nature* 385 (1997): 381.
Dixon, Bernard. "Books and Films: Powerful Media for Science Popularization." *Impact of Science on Society* 36.4 (1986): 379–85.
Douglas, Aileen. "Popular Science and the Representation of Women: Fontenelle and After." *Eighteenth-Century Life* 18.2 (1994): 1–14.
Doyle, Arthur Conan. *A Study in Scarlet*. 1887. London: Penguin, 1981.
Dundes, Alan ed. and introd. *Sacred Narrative: Readings in the Theory of Myth*. Berkeley: University of California Press, 1984.
Dunning-Davies, Jeremy. "Popular Status and Scientific Influence: Another Angle on 'The Hawking Phenomenon.'" *Public Understanding of Science* 2 (1993): 85–6.
Durant, John ed. *Museums and the Public Understanding of Science*. London: Science Museum, 1992.

———. Rev. of *The Faber Book of Science*, ed. John Carey. *Times Literary Supplement* 24 Nov. 1995: 5.

———. "What is Scientific Literacy?" *Science and Culture in Europe*. Ed. John Durant and Jane Gregory. London: Science Museum, 1993. 129–37.

Dyson, Freeman. "The Scientist as Rebel." *Nature's Imagination: The Frontiers of Scientific Vision*. Ed. John Cornwell. Oxford: Oxford University Press, 1995. 1–11.

Eagleton, Terry. *Literary Theory: An Introduction*. 1983. Oxford: Blackwell, 1995.

Eger, Martin. "Hermeneutics and the New Epic of Science." McRae, *Literature of Science* 186–209.

Ellmann, Lucy. "No Holes Barred." *Guardian* 25 June 1998, "Online": 1–3.

Enhager, Kjell. *Quantum Golf: The Path to Golf Mastery*. London: HarperCollins, 1991.

Evans, Dylan. "Hawking Started It." *Guardian* 10 Mar. 2005: 21.

Fahnestock, Jeanne. "Accommodating Science: The Rhetorical Life of Scientific Facts." McRae, *Literature of Science* 17–36.

———. "Preserving the Figure: Consistency in the Presentation of Scientific Arguments." *Written Communication* 21.1 (2004): 6–31.

Feyerabend, Paul. "Has the Scientific View of the World a Special Status Compared with Other Views?" *Physics and Our View of the World*. Ed. J. Hilgevoord. Cambridge: Cambridge University Press, 1994. 135–48.

Frayn, Michael. *Copenhagen*. London: Methuen Drama, 1998.

Friedman, Alan J. and Carol C. Donley. *Einstein as Myth and Muse*. Cambridge: Cambridge University Press, 1985.

Frost, William P. *What is the New Age? Defining Third Millennium Consciousness*. Lewiston, NY: Meller, 1992.

Froula, Christine. "Quantum Physics/Postmodern Metaphysics: The Nature of Jacques Derrida." *Western Humanities Review* 39.4 (1985): 287–313.

Fuller, Gillian. "Cultivating Science: Negotiating Discourse in the Popular Texts of Stephen Jay Gould." *Reading Science: Critical and Functional Perspectives on Discourses of Science*. London: Routledge, 1998. 35–62.

Fuller, Steve. "The Human Touch." *Independent on Sunday* 28 June 1998, "Sunday Review": 71.

———. *Science*. Buckingham: Open University Press, 1997.

Fyfe, Aileen. *Science and Salvation. Evangelical Popular Science Publishing in Victorian Britain*. Chicago: University of Chicago Press, 2004.

Gardner, Howard. *Creating Minds: An Anatomy of Creativity Seen Through the Lives of Freud, Einstein, Picasso, Stravinsky, Eliot, Graham, and Gandhi*. New York: BasicBooks-HarperCollins, 1993.

Gardner, Joseph H. "A Huxley Essay as 'Poem.'" *Victorian Studies* 14.2 (1970): 177–91.

Gardner, Martin. *The New Age: Notes of a Fringe Watcher*. Buffalo, NY: Prometheus, 1988.

Gartner, Carol B. "When Science Writing Becomes Literary Art: The Success of *Silent Spring*." Waddell 103–25.

Gates, Barbara T. "Retelling the Story of Science: Women Popularizers in Nineteenth-Century Britain." *Victorian Literature and Culture* 21 (1993): 289–306.

———, and Ann B. Shteir. *Natural Eloquence: Women Reinscribe Science*. Madison: University of Wisconsin Press, 1997.
Gieryn, Thomas F. *Cultural Boundaries of Science: Credibility on the Line*. Chicago: University of Chicago Press, 1999.
Gilder, George F. *Microcosm: The Quantum Revolution in Economics and Technology*. New York: Simon, 1989.
Ginsparg, Paul and Sheldon Glashow. "Desperately Seeking Superstrings?" *Physics Today* 39.5 (1986): 7–9.
Glick, Thomas F. ed. *The Comparative Reception of Relativity*. Boston Studies in the Philosophy of Science 103. Dordrecht: Reidel, 1987.
———. "Cultural Issues in the Reception of Relativity." Glick, *Comparative Reception* 381–400.
Gossin, Pamela. "Literature and the Modern Physical Sciences." *The Modern Physical and Mathematical Sciences*. Ed. Mary Jo Nye. Cambridge: Cambridge University Press, 2003. Vol. 5 of *The Cambridge History of Science*. David C. Lindberg and Ronald L. Numbers, gen. eds. 8 vols. 2001–. 91–109.
Gould, Stephen Jay. "Bright Star Among Billions." Editorial. *Science* 275 (1997): 599.
———. "Fulfilling the Spandrels of World and Mind." Selzer 310–36.
Greenberg, Valerie D. "The Scientific Text as Literary Artifact: Reading Max Planck." *New Orleans Review* 18.1 (1991): 56–63.
Gregory, Jane and Steve Miller. *Science in the Public: Communication, Culture and Credibility*. New York: Plenum, 1998.
Gross, Paul R. and Norman Levitt. *Higher Superstition: The Academic Left and its Quarrels with Science*. Baltimore: Johns Hopkins University Press, 1994.
———, Norman Levitt and Martin W. Lewis. *The Flight from Science and Reason*. Annals of the New York Academy of Science 775. New York: New York Academy of Science, 1996.
Hamilton, Alex. "Top Hundred Chart of 1996 Paperback Fastsellers." *Writers' and Artists' Yearbook 1998: Ninety-First Year of Issue*. London: Black, 1997. 261–8.
Hammett, Dashiell. *Red Harvest*. *Dashiell Hammett: The Four Great Novels*. London: Picador, 1982. 5–192.
Harris, Jack. "Popular Science: Its Backlash and the God Factor." *British Book News* Aug. 1992: 508–12.
Harris, R. Allen. "Rhetoric of Science." *College English* 53.3 (1991): 282–307.
Hayles, N. Katherine. *Chaos Bound: Orderly Disorder in Contemporary Literature and Science*. Ithaca: Cornell University Press, 1990.
———. "Constrained Constructivism: Locating Scientific Inquiry in the Theater of Representation." *New Orleans Review* 18.1 (1991): 76–85.
———. "Deciphering the Rules of Unruly Disciplines: A Modest Proposal for Literature and Science." Bruce and Purdy 25–48.
———. "Introduction: Complex Dynamics in Literature and Science." *Chaos and Order: Complex Dynamics in Literature and Science*. Ed. N. Katherine Hayles. Chicago: University of Chicago Press, 1991. 1–33.
———. "Turbulence in Literature and Science." *American Literature and Science*. Ed. Robert J. Scholnick. Lexington: Univeristy Press of Kentucky, 1992. 229–50.

Haynes, Roslynn. *From Faust to Strangelove: Representations of the Scientist in Western Literature.* Baltimore: Johns Hopkins University Press, 1994.

Heimann, P. M. "The Unseen Universe: Physics and the Philosophy of Nature in Victorian Britain." *British Journal for the History of Science* 6.21 (1972): 73–9.

Herbert, Roy. "How Do You Like Your Fudge?" *New Scientist* 18 Nov. 1995: 62.

Herman, David. "Silence of the Critics." *Prospect* Dec. 2002: 38–41.

Hess, David J. *Science and Technology in a Multicultural World: The Cultural Politics of Facts and Artifacts.* New York: Columbia University Press, 1995.

Hesse, Mary. "Cosmology as Myth." *Concilium* 166 (1983): 49–54.

———. *Revolutions and Reconstructions in the Philosophy of Science.* Brighton: Harvester, 1980.

Highfield, Roger. Personal interview. 15 March 1996.

———. Telephone interview. 27 February 1996.

Higley, Sarah L. "Alien Intellect and the Robiticization of the Scientist." *Camera Obscura* 40–41 (177): 131–62.

Hilgartner, Stephen. "The Dominant View of Popularization: Conceptual Problems, Political Uses." *Social Studies of Science* 20.3 (1990): 519–39.

Holman, Tim. Letter to the author. 1 Mar. 1999.

Honko, Lauri. "The Problem of Defining Myth." Dundes 41–52.

Hospital, Janette Turner. *Charades.* St. Lucia: University of Queensland Press, 1988.

Hudson, Reggie L. "Theory, Hypothesis and Sherlock Holmes." *Baker Street Journal* 41.2 (1991): 86–92.

"In Time with Hawking." Editorial. *Times* 3 July 1992: 15.

Jenkins, Helen. "On Being Clear About Time: An Analysis of a Chapter of Stephen Hawking's *A Brief History of Time*." *Language Sciences* 14 (1992): 529–44.

Johnson, Alex. Letter. *Guardian* 2 July 1998, "Online": 6.

Johnson, D. Barton. "The Ambidextrous Universe of Nabokov's *Look at the Harlequins!*" *Critical Essays on Vladimir Nabokov.* Ed. Phyllis A. Roth. Boston: Hall, 1984. 202–15.

Jung, C. G. and W. Pauli. *The Interpretation of Nature and the Psyche.* Trans. R. F. C. Hull and P. Silz. London: Routledge, 1955.

Jurdant, Baudouin. "Popularization of Science as the Autobiography of Science." *Public Understanding of Science* 2 (1993): 365–73.

Justice, Keith L. *Bestseller Index: All Books, by Author, on the Lists of Publishers Weekly and the New York Times Through 1990.* Jefferson, NC: McFarland, 1998.

Kellert, Stephen H. "Extrascientific Uses of Physics: The Case of Nonlinear Dynamics and Legal Theory." *Philosophy of Science* 68.3 Supplement (2001): S455-S466.

Kelley, Robert T. "Chaos Out of Order: The Writerly Discourse of Semipopular Scientific Texts." McRae, *Literature of Science* 132–51.

Kelly, Alfred. *The Descent of Darwin: The Popularization of Darwinism in Germany, 1860–1914.* Chapel Hill, NC: University of North Carolina Press, 1981.

Kemnitz, Charles. "Beyond the Zone of the Middle Dimensions: A Relativistic Reading of *The Third Policeman*." *Irish University Review* 15.1 (1985): 56–72.

Kenton, Branton. *Quantum Carrot: A New Concept in Small Space Organic Gardening.* London: Ebury, 1987.

Kenward, Michael. "The Other Scientific Literature." *New Scientist* 16 June 1988: 77.

———. "A Prize for Plain Speaking About Science." *New Scientist* 12 May 1990: 67–8.

Kevles, Daniel J. *The Physicists: The History of a Scientific Community in Modern America.* New York: Vintage-Random, 1979.

Kitcher, Philip. "A Plea for Science Studies." *A House Built on Sand: Exposing Postmodernist Myths About Science.* Ed. Noretta Koertge. 1998. New York: Oxford University Press, 2000. 32–56.

Knight, David M. *Natural Science Books in English 1600–1900.* New York: Praeger, 1972.

———. "Scientists and their Publics: Popularization of Science in the Nineteenth Century." *The Modern Physical and Mathematical Sciences.* Ed. Mary Jo Nye. Cambridge: Cambridge University Press, 2003. vol. 5 of *The Cambridge History of Science.* David C. Lindberg and Ronald L. Numbers, gen. eds. 8 vols. 2001–. 72–90.

Knight, Stephen. *Form and Ideology in Crime Fiction.* London: Macmillan, 1980.

Knorr-Cetina, Karin D. "The Constructivist Programme in the Sociology of Science: Retreats or Advances?" *Social Studies of Science* 12.2 (1982): 279–97.

———. "The Ethnographic Study of Scientific Work: Towards a Constructivist Interpretation of Science." Knorr-Cetina and Mulkay, *Science Observed* 115–40.

———, Karin D. and Michael Mulkay, eds. and introd. *Science Observed: Perspectives on the Social Study of Science.* London: Sage, 1983.

Knudsen, Susanne. "Scientific Metaphors Going Public." *Journal of Pragmatics* 35 (2003): 1247–63.

Koestler, Arthur. *The Roots of Coincidence.* London: Hutchison, 1972.

Kuhn, Thomas. *The Structure of Scientific Revolutions.* 1962. 2nd edn. Chicago: University of Chicago Press, 1970.

———. "Metaphor in Science." Ortony, *Metaphor and Thought* 533–42.

Kuritz, Hyman. "The Popularization of Science in Nineteenth-Century America." *History of Education Quarterly* 21.3 (1981): 259–74.

Labinger, Jay A. "The Science Wars and the Future of the American Academic Profession." *Daedalus* 126.4 (1997): 201–220.

——— and Harry Collins. Introduction. Labinger and Collins, *The One Culture?* 1–10.

———, eds. *The One Culture? A Conversation about Science.* Chicago: Univerisity of Chicago Press, 2001.

———. Preface. Labinger and Collins, *The One Culture?* ix–xi.

LaFollette, Marcel C. *Making Science Our Own: Public Images of Science 1910–1955.* Chicago: University of Chicago Press, 1990.

Latour, Bruno. "Visualization and Cognition: Thinking with Eyes and Hands." *Knowledge and Society: Studies in the Sociology of Culture Past and Present* 6 (1986): 1–40.

Leane, Elizabeth. "Reading Aldiss and Penrose's *White Mars* as Science Faction." *Foundation: The International Review of Science Fiction* 93 (2005): 18–25.

Lee, Judith Yaross. Introduction. Lee and Slade 75–7.

———, and Joseph W. Slade, eds. *Beyond the Two Cultures: Essays on Science, Technology and Literature*. Ames, IA: Iowa University Press, 1990.

Leitch, Alan. Presentation delivered as part of the British Association Annual Festival of Science Seminar: "The Book, the Niche, the Strife and the Cover: Popular Science Writing in the 1990s." 11 Sept. 1996.

LeShan, Lawrence. *The Medium, The Mystic, and the Physicist: Towards a General Theory of the Paranormal.* 1974. New York: Penguin-Arkana, 1995.

Lessl, Thomas M. "Science and the Sacred Cosmos: The Ideological Rhetoric of Carl Sagan." *Quarterly Journal of Speech* 71 (1985): 175–87.

———. "The Priestly Voice." *Quarterly Journal of Speech* 75 (1989): 183–97.

Levin, Bernard. "Brave Face of a Life on Tick." *Times* 22 Apr. 1991: 12.

Levine, George, ed. *One Culture: Essays in Science and Literature*. Madison: University of Wisconsin Press, 1987.

———. "One Culture: Science and Literature." Levine, *One Culture* 3–32.

Lewenstein, Bruce. "How Science Books Drive Public Discussion." *Communicating the Future: Best Practices for Communication of Science and Technology to the Public*. Washington: GPO, 2002. 69–76.

———. "Was There Really a Popular Science 'Boom'?" *Science, Technology, and Human Values* 12.2 (1987): 29–41.

———. "Why Isn't Popular Science More Popular?" *American Scientist* 76 (1988): 447–9.

Lewontin, R. C. Steven Rose, and Leon J. Kamin. *Not in Our Genes: Biology, Ideology, and Human Nature*. New York: Pantheon-Random, 1984.

Lightman, Bernard. "The Story of Nature: Victorian Popularizers and Scientific Narrative." *Victorian Review* 25.2 (1999): 6–29.

———. "'The Voices of Nature': Popularizing Victorian Science." *Victorian Science in Context*. Ed. Bernard Lightman. Chicago: University of Chicago Press, 1997. 187–211.

Lincoln, Bruce. *Theorizing Myth: Narrative, Ideology, and Scholarship*. Chicago: University of Chicago Press, 1999.

Luey, Beth. "'Leading the Public Gently': Popular Science Books in the 1950s." *Book History* 2.1 (1999): 218–53.

Lyotard, Jean-François. *The Postmodern Condition: A Report on Knowledge*. Trans. Geoff Bennington and Brian Massumi. Theory and History of Literature 10. Manchester: Manchester University Press, 1989.

Macdonald-Ross, Michael. "The Role of Science Books for the Public." *Communicating Science to the Public*. Ed. David Evered and Maeve O'Connor. Proc. of Ciba Foundation Conference, Oct. 1986, London. Chicester: Wiley, 1987. 175–89.

Macilwain, Colin. "'Science and Reason' Forum Finds Enemies All Around." *Nature* 375 (1995): 439.

Maddox, John. "The Big Big Bang Book." *Nature* 336 (1988): 267.

March, Robert H. Rev. of *The Dancing Wu Li Masters*, by Gary Zukav. *Physics Today* 32.8 (1979): 54–5.

Markley, Robert. "What Now? An Introduction to Interphysics." *New Orleans Review* 18.1 (1991): 5–8.

McEwan, Ian. *The Child in Time.* London: Cape, 1987.
———. *Enduring Love.* London: Cape, 1997.
McRae, Murdo William ed. *The Literature of Science: Perspectives on Popular Scientific Writing.* Athens: University of Georgia Press, 1993.
———. "Introduction: Science in Culture." McRae, *Literature of Science* 1–13.
Meadows, A. J. and M. M. Hancock, comps. and introd. *Front Page Physics: A Century of Physics in the News.* Bristol: Institute of Physics, 1994.
Medawar, Peter. "Science and Literature." *The Hope of Progress.* London: Methuen, 1972. 18–38.
———. "Is The Scientific Paper a Fraud?" 1963. *The Threat and the Glory: Reflections on Science and Scientists.* Ed. David Pyke. Oxford: Oxford University Press, 1990. 228–33.
Mellor, Felicity. "Between Fact and Fiction: Demarcating Science from Non-Science in Popular Physics Books." *Social Studies of Science* 33.4 (Aug. 2003): 509–538.
Mermin, N. David. *Boojums All The Way Through: Communicating Science in a Prosaic Age.* New York: Cambridge University Press, 1990.
———. Letter. *Physics Today* 49.7 (1996): 13–15.
Mialet, Helen. "Do Angels Have Bodies? Two Stories About Subjectivity in Science: The Cases of William X and Mister H." *Social Studies of Science* 29.4 (1999): 551–81.
———. "Reading Hawking's Presence: An Interview with a Self-Effacing Man." *Critical Inquiry* 29.4 (2003): 571–98.
Michael, John. "Prosthetic Gender and Universal Intellect: Stephen Hawking's Law." *Boys: Masculinities in Contemporary Culture.* Ed. P. Smith. Boulder, CO: Westview, 1996. 199–218.
Midgley, Mary. *Science as Salvation: A Modern Myth and its Meaning.* London: Routledge, 1992.
Millhauser, Milton. "Dr. Newton and Mr. Hyde: Scientists in Fiction from Swift to Stevenson." *Nineteenth-Century Fiction* 28.3 (1973): 287–304.
Mirchandani, Ravi. "The Art of the Pop Picker." *New Scientist* 4 June 1994: 42.
———. Telephone interview. 26 February 1996.
"Mistletoe Cutters." Rev. of *Superforce: The Search for a Grand Unified Theory of Nature*, by Paul Davies. *The Economist* 17 Nov. 1984: 97.
Moirland, Sophie. "Communicative and Cognitive Dimensions of Discourse on Science in the French Mass Media." *Discourse Studies* 5.2 (2003): 175–206.
Montgomery, Scott. *The Scientific Voice.* New York: Guilford, 1996.
Myers, Greg. "Discourse Studies of Scientific Popularization: Questioning the Boundaries." *Discourse Studies* 5.2 (2003): 265–79.
———. "Fictionality, Demonstration, and a Forum for Popular Science: Jane Marcet's *Conversations on Chemistry*." Gates and Shteir 43–60.
———. "Fictions for Facts: The Form and Authority of the Scientific Dialogue." *History of Science* 30 (1992): 221–47.
———. "Making a Discovery: Narratives of Split Genes." *Narrative in Culture: The Uses of Storytelling in the Sciences, Philosophy, and Literature.* Ed. Christopher Nash. London: Routledge, 1990. 102–126.
———. "Nineteenth-Century Popularizations of Thermodynamics and the Rhetoric of Social Prophecy." *Victorian Studies* 29 (1985): 35–66.

———. "The Pragmatics of Politeness in Scientific Texts." *Applied Linguistics* 10 (1989): 1–35.

———. "Science for Women and Children: The Dialogue of Popular Science in the Nineteenth Century." Christie and Shuttleworth 171–200.

———. *Writing Biology: Texts in the Social Construction of Scientific Knowledge*. Madison, WI: Univeristy of Wisconsin Press, 1990.

Nieman, Adam. "The Popularisation of Physics: Boundaries of Authority and the Visual Culture of Science." Diss. University of West England, Bristol, 2000.

Nixon, Will. "The Art of Publishing Popular Science Books." *Publishers Weekly* 23 Aug. 1991: 32–5.

Norris, Christopher. "Quantum Worlds Without End: On David Deutsch's *The Fabric of Reality*." *Southern Humanities Review* 33.2 (1999): 149–81.

Nye, Mary Jo. "Introduction: The Modern Physical and Mathematical Sciences." *The Modern Physical and Mathematical Sciences*. Ed. Mary Jo Nye. Cambridge: Cambridge University Press, 2003. Vol. 5 of *The Cambridge History of Science*. David C. Lindberg and Ronald L. Numbers, gen. eds. 8 vols. 2001–. 1–17.

Nyman, Jopi. *Men Alone: Masculinity, Individualism, and Hard-Boiled Fiction*. Amsterdam: Rodopi, 1997.

Okuda, Michael, Denise Okuda and Debbie Mirek. *The Star Trek Encyclopedia: A Reference Guide to the Future*. New York: Pocket-Simon, 1994.

Oldershaw, Robert L. "The New Physics—Physical or Mathematical Science?" *American Journal of Physics* 56 (1988): 1075–81.

———. "What's Wrong with the New Physics?" *New Scientist* 22/29 Dec. 1990: 56–9.

Ortony, Andrew, ed. *Metaphor and Thought*. 2nd edn. Cambridge: Cambridge University Press, 1996.

———. "Metaphor, Language, and Thought." Ortony, *Metaphor and Thought* 1–16.

The Oxford English Dictionary. 2nd edn. Oxford: Oxford University Press, 1989.

Parker, Robert B. *The Private Eye in Hammett and Chandler*. Northridge, CA: Lord John, 1984.

Parrinder, Patrick. "Scientists in Science Fiction: Enlightenment and After." *Science Fiction Roots and Branches: Contemporary Critical Approaches*. Ed. Rhys Garnett and R. J. Ellis. New York: St Martin's, 1990. 57–78.

Paul, Danette. "Spreading Chaos: The Role of Popularizations in the Diffusion of Scientific Ideas." *Written Communication* 21.1 (2004): 32–68.

Paulson, William R. *The Noise of Culture: Literary Texts in a World of Information*. Ithaca: Cornell University Press, 1988.

Perkowitz, Sydney. "Real Physicists Don't Wear Ties." *New Scientist* 21 Dec. 1991: 22–4.

Peterfreund, Stuart. Introduction. *Literature and Science: Theory and Practice*. Ed. Stuart Peterfreund. Boston: Northeastern University Press, 1990. 3–12.

Petkova, Kristina and Pepka Boyadjieva. "The Image of the Scientist and its Functions." *Public Understanding of Science* 3 (1994): 215–24.

Porlock, Harvey. "Critical List." *Sunday Times* 8 Sept. 1996, sec. 7 ("Books"): 2.

Prelli, Lawrence J. *A Rhetoric of Science: Inventing Scientific Discourse*. Columbia: University of South Carolina Press, 1989.

Quantum Leap. NBC. 26 March 1989–5 May 1993.

Quine, W. V. *From a Logical Point of View: Nine Logico-Philosophical Essays.* Rev. edn. Cambridge, MA: Harvard University Press, 1980.

Rahm, Jrène and Paul Charbonneau. "Probing Stereotypes Through Students' Drawings of Scientists." *American Journal of Physics* 65.8 (1997): 774–8.

Restivo, Sal. "Parallels and Paradoxes in Modern Physics and Eastern Mysticism: I—A Critical Reconnaissance." *Social Studies of Science* 8.2 (1978): 143–81.

——. "Science Studies—What Is to Be Done." *Science, Technology, and Human Values* 12.2 (1987): 13–18.

Richards, I. A. *The Philosophy of Rhetoric.* New York: Oxford University Press, 1936.

Robbins, Bruce and Andrew Ross. Editorial response to Sokal. In "Mystery Science Theater." *Lingua Franca* 6.5 (1996): 54–64. 2 Oct. 2004. <http://www.geocities.com/scirevolution/socialtext.html>.

Rodgers, Michael. "The Hawking Phenomenon." *Public Understanding of Science* 1 (1992): 231–4.

Rogers, Moira R. *Newtonianism for the Ladies and Other Uneducated Souls: The Popularization of Science in Leipzig, 1687–1750.* NY: Peter Lang, 2003.

Rose, Hilary. "My Enemy's Enemy Is—Only Perhaps—My Friend." Ross, *Science Wars* 80–101.

Rose, Phyllis. "Huxley, Holmes, and the Scientist as Aesthete." *Victorian Newsletter* 38 (1970): 22–4.

Rose, Steven, R. C. Lewontin and Leon J. Kamin. *Not in Our Genes: Biology, Ideology and Human Nature.* Harmondsworth: Penguin, 1984.

Rosenheim, Shawn. "Extraterrestrial: Science Fictions in 'A Brief History of Time' and 'The Incredible Shrinking Man.'" *Film Quarterly* 48.4 (1995): 15–21.

Ross, Andrew. "Burden of Spoof." *Times Higher Education Supplement* 21 June 1996: 16.

—— ed. *Science Wars.* Durham, NC: Duke University Press, 1996.

——. *Strange Weather: Culture, Science and Technology in an Age of Limits.* London: Verso, 1991.

Rousseau, G. S. "Science Books and their Readers in the Eighteenth Century." *Books and their Readers in Eighteenth-Century England.* Ed. Isabel Rivers. Leicester: Leicester University Press, 1982. 197–255.

——. "Virtual Realities." Rev. of *From Faust to Strangelove: Representations of the Scientist in Western Literature*, by Roslynn D. Haynes, *Realism and Representation: Essays on the Problem of Realism in Relation to Science*, ed. George Levine, and *Life on the Screen: Identity in the Age of the Internet*, by Sherry Turkle. *British Journal for the History of Science* 30 (1997): 227–32.

Russell, Bertrand. *Mysticism and Logic.* 2nd edn. London: Unwin, 1963. Rpt. of *Philosophical Essays.* 1910.

Schachterle, Lance. "The Metaphorical Allure of Modern Physics: Introduction." Lee and Slade 177–84.

Schatzberg, Walter, Ronald A. Waite, Jonathan K. Johnson, eds. *The Relations of Literature and Science: An Annotated Bibliography of Scholarship, 1880–1980.* New York: MLA, 1987.

Schickore, Jutta. "The Task of Explaining Sight——Helmholtz's Writings on Vision as a Test Case for Models of Science Popularization." *Science in Context* 14 (2001): 397–417.

Schmidt, Siegfried J. "The Logic of Observation: An Introduction to Constructivism." *Canadian Review of Comparative Literature* 19.3 (1992): 295–311.

Schwartz, A. Truman. "Some Unsolicited Advice to Popularizers (and Teachers) of Science." *Journal of Chemical Education* 67 (1990): 754–6.

"Science Wars and the Need for Respect and Rigour." Opinion. *Nature* 385 (1997): 373.

Searle, John R. "The World Turned Upside Down." Rev. of *On Deconstruction: Theory and Criticism after Structuralism*, by Jonathon Culler. *New York Review of Books* 27 Oct. 1983: 74–9.

Segal, Robert A. ed. "Series Introduction." *Theories of Myth: From Ancient Israel and Greece to Freud, Jung and Campbell.* 6 vols. Vol. 2. *Literary Criticism and Myth*. New York: Garland, 1996. vii–xi.

Sehgal, Narender K., Satpal Sangwan and Subodh Mahanti. *Uncharted Terrains: Essays on Science Popularisation in Pre-Independence India*. New Delhi: Vigyan Prasar, 2000.

Selzer, Jack ed. *Understanding Scientific Prose*. Madison: University of Wisconsin Press, 1993.

Shaffer, Elinor S. Introduction. *The Third Culture: Literature and Science*. Ed. Elinor S. Shaffer. European Culture: Studies in Literature and the Arts 9. Berlin: de Gruyter, 1998.

Sheets-Pyenson, Susan. "Popular Science Periodicals in Paris and London: The Emergence of a Low Scientific Culture, 1820–1875." *Annals of Science* 42 (1985): 549–72.

Shermer, Michael B. "This View of Science: Stephen Jay Gould as Historian of Science and Scientific Historian, Popular Scientist and Scientific Popularizer." *Social Studies of Science* 32.4 (2002): 489–524.

Shinn, Terry and Richard Whitley eds. *Expository Science: Forms and Functions of Popularisation*. Sociology of the Sciences Yearbook 9. Dordrecht: Reidel, 1985.

———. Preface. Shinn and Whitley vii–ix.

Shteir, Ann. "Botanical Dialogues: Maria Jacson and Women's Popular Science Writing in England." *Eighteenth-Century Studies* 23.3 (1990): 301–17.

Sigman, Joseph. "Science and Parody in Kurt Vonnegut's *The Sirens of Titan*." *Mosaic* 19.1 (1986): 15–32.

Smith, Roger. *Popular Physics and Astronomy: An Annotated Bibliography*. Lanham, MD: Scarecrow, 1996.

Snow, C. P. *The Two Cultures*. 1959, 1964. Cambridge: Canto-Cambridge University Press, 1993.

Sokal, Alan. "A Physicist Experiments with Cultural Studies." *Lingua Franca* 6.4 May/June 1996: 62–4. 5 Oct. 2004. <http://www.physics.nyu.edu/faculty/sokal/lingua_franca_v4/lingua_franca_v4.html>.

———. "Transgressing the Boundaries: An Afterword." *Dissent* 43.4 (1996): 93–9. Rpt. in Sokal and Bricmont 248–58.

———. "Transgressing the Boundaries: Towards a Transformative Hermeneutics of Quantum Gravity." *Social Text* 46/47 (1996): 217–52. Rpt. in Sokal and Bricont 199–240.

——— and Jean Bricmont. *Intellectual Impostures: Postmodern Philosophers' Abuse of Science*. London: Profile, 1998.

Sokolov, Raymond A. Rev. of *The Dancing Wu Li Masters*, by Gary Zukav. *New York Times Book Review* 17 June 1979: 18.
Sporn, Paul. "The Modern Physics of Contemporary Criticism." Lee and Slade 201–22.
Sprat, Thomas. *History of the Royal Society*. 1667. Ed. Jackson I. Cope and Harold Whitmore Jones. St Louis: Washington University Press, 1959.
Stone, David A. "*Omni* Meets Feynman: The Interaction Between Popular and Scientific Cultures." McRae, *Literature of Science* 291–310.
Stoppard, Tom. *Arcadia*. London: Faber, 1993.
——. *Hapgood*. London: Faber, 1988.
Sulkin, Will. Personal interview. 22 Oct. 1996.
Sullivan, J. W. N. "The Entente Cordiale." *Athenæum* 9 Apr. 1920: 482.
——. "Popular Science." *Athenæum* 1 Oct. 1920: 444–5. Rpt. in *Aspects of Science*. London: Cobden-Sanderson, 1923. 82–8.
——. *Three Men Discuss Relativity*. London: Collins, 1925.
Tallack, Peter. "Echo of the Big Bang: An End to the Boom in Popular Science Books May Actually Raise Standards." *Nature* 432 (2004): 803–804.
Terzian, Yervant and Elizabeth Bilson eds. Preface. *Carl Sagan's Universe*. Cambridge: Cambridge University Press, 1997. xi–xii.
Toulmin, Stephen. "Scientific Mythology." *The Return to Cosmology: Postmodern Science and the Theology of Nature*. Berkeley: University of California Press, 1982. 1–85. Rpt. of "Contemporary Scientific Mythology." *Metaphysical Beliefs*. By Stephen E. Toulmin, Donald W. Hepburn and Alasdair MacIntyre. 1957. London: SCM: 1970. 3–71.
Tudge, Colin. "In Praise of Science Writers." *New Scientist* 20 Nov. 1986: 44–8.
——. "Let Us Into the Tower of Knowledge." *Guardian* 13 Mar. 1999, "Saturday Review": 3.
Turner, Frank M. "Public Science in Britain, 1880–1919." *Isis* 71.4 (1990): 589–608.
Turner, Jenny. "Scientific Sex Appeal." *Vogue* Apr. 1997: 40–43.
Turney, Jon. "Accounting for Explanation in Popular Science Texts: An Analysis of Popularized Accounts of Superstring Theory." *Public Understanding of Science* 13.4 (2004): 331–46.
——. "More Than Story-Telling: Reflecting on Popular Science." *Science Communication in Theory and Practice*. Ed. Susan M. Stocklmayer, Michael M. Gore and Chris Bryant. Dordrecht: Kluwer Academic, 2001. 47–62.
——. "Natural Selection." *Guardian* 15 Dec. 2001, "Guardian Unlimited Books". 19 May 2006. <http://books.guardian.co.uk/bestof2001/story/0,,620212,00.html>.
——. "New Instant Science!" *Times Higher Education Supplement* 10 May 1996: 15.
——. "Telling the Facts of Life: Cosmology and the Epic of Evolution." *Science as Culture* 10.2 (2001): 225–47.
——. "The Word and the World: Engaging with Science in Print." *Communicating Science: Context and Channels*. London: Routledge, 1999. 120–33.
"Unscientific Readers of Science." *Economist* 9 May 1998: 129–30.

van Dijck, José. "After the 'Two Cultures': Towards a '(Multi)cultural' Practice of Science Communication." *Science Communication* 25.2 (2003): 177–90.
Vanderbeke, Dirk. "Physics, Rhetoric, and the Language of *Finnegans Wake*." *The Languages of Joyce: Selected Papers from the 11th International James Joyce Symposium, Venice, 12–18 June 1988*. Ed. R. M. Bollettieri Bosinelli, C. Marengo Vaglio, and Christine van Boheemen. Philadelphia: Benjamins, 1992. 249–56.
Vines, Gail. "The Cheque's in the Post." *New Scientist* 26 July 1997: 58–9.
Waddell, Craig, ed. *And No Birds Sing: Rhetorical Analyses of Rachel Carson's Silent Spring*. Carbondale, IL: Southern Illinois University Press, 2000.
Wasserstein, Alan G. "Aggression and Power: The R-complex and Nuclear Blackmail." McRae, *Literature of Science* 230–48.
Weart, Spencer. "The Physicist as Mad Scientist." *Physics Today* 41.6 (1988): 28–37.
Weber, Renée. *Dialogues with Scientists and Sages: The Search for Unity*. London: Penguin, 1986.
Weinberg, Steven. *Facing Up: Science and its Cultural Adversaries*. Cambridge, MA: Harvard University Press, 2001.
———. "Peace at Last?" Labinger and Collins, *The One Culture?* 238–40.
Weininger, Stephen J. "Afterword." *New Orleans Review* 18.1 (1991): 100–103.
Weldon, Fay. "Thoughts We Dare Not Speak Aloud." *Daily Telegraph* 2 Dec. 1991: 14.
Wertheim, Margaret. *Pythagoras' Trousers: God, Physics, and the Gender Wars*. London: Fourth Estate, 1997.
Wetzels, Walter D. "The Popularisation of a New Physics: The Case of Leonhard Euler's *Letters to a German Princess*." *Studies on Voltaire and the Eighteenth Century* 264 (1989): 796–800.
What the Bleep Do We Know!? Dir. William Arntz, Betsy Chasse and Mark Vincent. Samuel Goldwyn Films and Roadside Attractions, 2004.
White, Eric. "Contemporary Cosmology and Narrative Theory." *Literature and Science: Theory and Practice*. Ed. Stuart Peterfreund. Boston: Northeastern University Press, 1990. 91–112.
White, Hayden. "The Value of Narrativity in the Representation of Reality." *On Narrative*. Ed. W. J. T. Mitchell. Chicago: University of Chicago Press, 1981. 1–23.
White, Michael. "Making the Quantum Leap." *Sunday Times* 19 Nov. 1995, sec. 7 ("Books"): 16.
——— and John Gribbin. *Stephen Hawking: A Life in Science*. London: Penguin, 1992.
Whitley, Richard. "Knowledge Producers and Knowledge Acquirers: Popularisation as a Relation Between Scientific Fields and Their Publics." Shinn and Whitley 3–28.
Whitworth, Michael H. "The Clothbound Universe: Popular Physics Books, 1919–1939." *Publishing History* 40 (1996): 53–82.
———. *Einstein's Wake: Relativity, Metaphor, and Modernist Literature*. Oxford: Oxford University Press, 2001.
———. "Physics and the Literary Community, 1905–1939." Diss. University of Oxford, 1994.
Wilber, Ken ed. *Quantum Questions: Mystical Writings of the World's Great Physicists*. Boulder, CO: Shambhala, 1984.

Wilder, Billy and Raymond Chandler, adapt. *Double Indemnity*. By James M. Cain. *Chandler: Later Novels and Other Writings*. By Raymond Chandler. New York: Library of America, 1995. 873–972.

Wilson, David B. "The Thought of Late Victorian Physicists: Oliver Lodge's Ethereal Body." *Victorian Studies* 15.1 (1971): 29–48.

Wolpert, Lewis. "So Much For Artistic Licence." *Daily Telegraph* 9 Dec. 1991: 12.

——. *The Unnatural Nature of Science*. London: Faber, 1992.

—— and Anthony Gottlieb. "Debate: Science Against Philosophy." *Prospect* Feb. 1997: 16–19.

—— and Alison Richards. "The Insiders' Story." *Independent on Sunday* 21 Sept. 1997, "The Sunday Review": 44–45.

—— and Alison Richards. *Passionate Minds: The Inner World of Scientists*. Oxford: Oxford University Press, 1997.

Woolgar, Stephen. *Science: The Very Idea*. Chichester: Ellis Horwood, 1988.

——. "What is a Scientific Author?" *What is an Author?* Ed. Maurice Biriotti and Nicola Miller. Manchester: Manchester University Press, 1993. 175–90.

Wynne, Brian. "Physics and Psychics: Science, Symbolic Action, and Social Control in Late Victorian England." *Natural Order: Historic Studies of Scientific Culture*. Ed. Barry Barnes and Steven Shapin. Beverly Hills: Sage, 1979. 167–86.

Yeats, W. B. "Among School Children." *Collected Poems*. 1933. Rev. edn. London: Pan, 1990. 242–5.

Index

Alfvén, Hannes
   on incomprehensibility encouraged by popularizers 24–5
   criticism of Big Bang cosmology 118, 120, 125
   predictions of research 125 (n10)
Amis, Martin
   popular science in work of 2–3, 36–7, 58
anthropomorphic metaphor
   in quantum physics popularizations 17, 28, 83, 88–94
   and "quantum consciousness" 93–5, 98–102
   in *Dancing Wu Li Masters, The* 95–105
Asimov, Isaac, 20, 28 (n7), 30, 165
   influence of 52
   use of mythic narrative by 121
astronomy, popularizations of 21, 22, 27, 30
   see also cosmology
atomic physics 29
   popularizations of 27–28
   see also quantum physics
atomic weapons 29, 30
Atwood, Margaret
   popular science in work of 2–3

Barrow, John
   criticism of 117–18
   popularizations by 36 (n11)
   recognition of narrative conventions by 121 (n8)
   and "two cultures debate" 55
Big Bang theory
   criticism of 118–19, 125–6, 164
   in science popularizations 12, 30, 35–38, 107–131, 135
   and "two cultures" debate 107, 115, 118
   see also Hawking, Stephen; Weinberg, Steven
black holes 30–31, 35, 93–4
Bohr, Niels 28, 105 (n17)
   Copenhagen interpretation of quantum theory 87, 93
   and Eastern philosophy 31
   and "quantum consciousness" 93
   use of metaphor by 87
Bragg, Melvyn
   sales of books by 45
   and science popularizations 2, 51
   and "two cultures" debate 55 (n5), 56–7
Bruce, Colin
   Strange Case of Mrs Hudson's Cat, The 145–6
   attack on constructivist theory in 75

Calder, Nigel
   Key to the Universe, The 21, 30–31, 132
   sales of books by 44
Capra, Fritjof 96, 100, 164
   Tao of Physics, The 32–3, 81, 95, 96, 99
   use of metaphor in 86, 97 (n10), 152
   used in literary work 2
Carroll, Lewis
   appropriation of work by science popularizers 87, 90–91
chaos theory 17
   definition of 139–40 (n6)
   popularizations of 38–9 *see also* Gleick, James
Chandler, Raymond 143, 144, 148, 149
characterization 138–39, 164, 165
   "absent–minded professor" 150–51, 153, 156, 157, 160, 165

detective fiction conventions 75, 140,
143–50, 157, 159, 160, 164–5
scientist as obsessive 153–4, 156, 160,
165
scientist as priest 151–5, 159, 165
scientist removed from social
conventions 154–7, 165
see also scientists, image of
Clarke, Arthur C. 30, 165
complexity theory 17
definition of 139–40 (n6)
popularizations of 38–39 *see also*
Waldrop, M. Mitchell
constructivist theories
relativist-constructivist 67, 69, 70, 71,
72, 74 (n3), 75, 77
of scientific knowledge 10, 14, 61, 65,
66–7, 68, 69, 70, 71, 72, 73, 115,
163
approach to language 83–84
attacks on 71–3, 76, 107, 115–16,
119, 135, 156 *see also* "Science
Wars"
definition of 65, 67, 68
cosmology 17
popularizations of 18, 27, 35–8
superstring theory 124–5
see also Big Bang theory; Hawking,
Stephen; Sagan, Carl; Weinberg,
Steven

Darwin, Charles 15, 129
Origin of Species, The
readership of 20, 23, 89
use of metaphor in 89–90, 92, 93,
102
Davies, Paul 30, 49, 50, 165
About Time 108
approach to characterization 139
attacks on constructivist theory by 72,
115
categorization of physics by 6, 17
Cosmic Blueprint 39
Forces of Nature, The
readership of 12 (n7)
God and the New Physics 37

sales of 47
influences in work of 30
Last Three Minutes, The
use of metaphor in 27
Matter Myth, The 39
Mind of God, The 38
on the publishing "boom" 42
Other Worlds 34, 92, 93, 95, 96
popularization used in literary work 2
and quantum theory popularizations
34, 39
reaction to *A Brief History of Time* 35,
48, 49, 50
rejection of anthropocentrism in
quantum theory 94
Superforce 33, 36
theological concerns of 37, 38
and "two cultures" debate 53, 55, 56, 58
use of gendered language by 158 (n15),
158–9
use of metaphors by 92, 93
Dawkins, Richard 50
attacks on constructivist theory by 72,
73, 75–6, 115
Blind Watchmaker, The 53
gendered language in 158
influence of 52
River Out of Eden
cross-disciplinary debate in 4
sales of books by 43, 45
and "two cultures" debate 55–6, 57
detective fiction conventions
character types in 148, 149, 160
use in science popularizations 75,
140–50, 157, 159, 164–5
Durant, John 5
on characterization of scientists 154–5
on criticism of science popularizations
5, 14
on heterogeneity of scientific culture
62–3
Dyson, Freeman
criticism of 117–18
influence of 52
on the humanity of scientists 155, 165

Eddington, Arthur 16, 22, 24, 28, 39, 56, 58
　Nature of the Physical World, The 21, 25, 27, 31, 87, 94
　Philosophy of Physical Science, The 25
　and "quantum consciousness" 94
　sales of books by 46
　Science and the Unseen World 23
　Space, Time and Gravitation 24
Einstein, Albert 97, 137, 146
　General Theory of Relativity 24, 118
　image of 150, 151–5, 156
　popularizations of theories 4, 24–5, 27–8, 31, 32, 34, 75, 91
　Special Theory of Relativity 19
　use of detective stereotype by 146, 148

Feynman, Richard P. 21, 30, 52, 55, 82, 112, 115, 117
　attitude to women 159
　Character of Physical Law, The 2
　image of 150–51
　influence of 52
　Meaning of It All, The 105
　　sales of 45
　　treatment of mythic narrative in 123
　QED: The Strange Theory of Light and Matter 92, 110
　on superstring theory 125 (n9)
　"*Surely You're Joking, Mr Feynman!*" 150, 151
　use of metaphor by 92, 102
　"*What Do You Care What Other People Think?*" 150
Fontenelle, Bernard le Bouvier de
　Entretiens sur la pluralité des mondes 20–21
Frayn, Michael 57
　popular science in work of 2–3

Gamow, George 16, 29–30
　Mr Tompkins Explores the Atom 28
　Mr Tompkins in Wonderland 28
　use of metaphor by 90
Gardner, Martin 30, 92
　Ambidextrous Universe, The 30
　Relativity for the Million 30

Gell-Mann, Murray 127, 141, 144
　popularizations by 39 (n16), 50
　publisher's advance for 49–50
　and "two cultures" debate 56
gender stereotypes 153 (n12), 157–60
General Theory of Relativity *see* Einstein, Albert
Gilmore, Robert
　Alice in Quantumland 90–91
Gleick, James 39
　Chaos 17, 38–9, 108, 140–41(n7)
　　characterization of scientists in 18, 140–50, 154, 159, 160–61, 164–5
　　female characters in 157, 158, 159
　　narrative style 147, 148
　　readership of 11
　　textual strategies in 38, 139–40, 141–3
　Genius: Richard Feynman and Modern Physics 39
　　image of Richard Feynman in 150–51
Gould, Stephen Jay 14, 74 (n3), 130, 138
　on constructivist theory 72
　influence of 52
　opinion of Carl Sagan 35
　and publishing "boom" 39, 43
　sales of books by 46, 47
　and "two cultures" debate 56, 57
Gribbin, John 39, 165
　approach to narrative 110–11
　approach to characterization 138–9
　Blinded by the Light 36
　denouncement of "New Age" movement 33
　In Search of Schrödinger's Cat 33, 34, 91–2, 95–6, 110
　influences on work of 30
　Matter Myth, The 39
　popularization used in literary work 2
　reaction to *A Brief History of Time* 49
　use of metaphors by 86, 91–2
Guth, Alan
　popularization by 50
　use of metaphor by 86

Hammett, Dashiell 143, 144, 148
Hawking, Stephen 11, 12, 30, 35, 36, 39, 45
   A Brief History of Time 5, 7, 17, 35, 37–8, 107, 110, 118 (n6)
      female characters in 158
      popular and critical reaction to 1, 35, 48–9, 117
      and publishing "boom" 42, 43, 47–51
      readership of 11, 12
      sales of 26, 42, 46, 47
      use of mythic narrative in 18, 107, 111–12, 114 (n5), 120, 122–3, 126, 127, 128, 129–131, 164
   image of 48, 54, 112, 113, 131, 132–4, 137 (n2), 138, 139, 152, 159
   influence of 52
   popularization used in literary work 2
   use of spirituality in popularizations 37–8, 152
   and "two cultures" debate 57
   writing style 48–9
Heisenberg, Werner 28, 105 (n17)
   Copenhagen interpretation of quantum theory 87
   Physics and Philosophy 28, 29, 87–8
   Uncertainty Principle 81, 93
   on use of metaphor 88
history of science popularizations 19
   physics popularizations 20–39, 42, 46, 47
Holmes, Sherlock, 143, 145–6, 148, 149, 160, 165
   appropriation by popularizers of 75, 145–6
Hospital, Janet Turner
   Charades 109–10, 119
   popular science in work of 2–3, 119
Hoyle, Fred 16, 29–30
Huxley, T. H. 24, 25, 146
   popularization by 22
   and "two cultures" debate 56
   writing style of 15

Jeans, James 16, 22, 28, 39, 56, 58
   Atomicity and Quanta 27

Mysterious Universe, The 7, 21, 23, 25, 27, 28, 31
   sales of 25, 26, 46
   textual strategies in 26–7
Physics and Philosophy 25

Kuhn, Thomas
   on heterogeneity of scientific culture 63
   theories of 64, 68, 168
      use in science popularizations 38
   on use of metaphor in popularizations 87, 109

Lederman, Leon 37
   on *Dancing Wu Li Masters, The* 95
   God Particle, The 110
      denouncement of "New Age" movement 33
      emphasis on aesthetics in 119
      parody in 38
      recognition of narrative conventions by 112, 115, 127
literary criticism
   theoretical approaches to science popularizations 69–70, 73–5
   post-structuralist 69, 70, 76–7, 164
literary critics
   reaction to science popularizations 4–5, 14–16, 51–8, 95–6, 104–106, 108–109, 164
   criticism of mythic narrative 116–17
   preference for "New Age" interpretations 95–6, 164
   reaction to publishing "boom" 42, 54, 55–6, 57–8
   and "two cultures" debate 61–4
literature
   inclusion of popular science in 2–3, 36–7, 54, 57, 58, 86, 107, 108, 109–110, 119
   representations of science in 3
   science popularizations and 1–2, 3–5, 58
   see also "two cultures" debate
Lodge, Oliver 31

Atoms and Rays 27
Ether and Reality 25
Lord Kelvin *see* Thomson, William

McEwan, Ian 57
   Enduring Love 107, 108, 119
   popular science in work of 2–3, 58, 119
Maxwell, James Clerk 21, 22
Medawar, Peter
   and science writing 65, 70, 108
   and "two cultures" debate 62
Mermin, David
   Boojums All The Way Through 4
   on scientific writing 138
   on theories of science 71
metaphor 17–18, 26, 27
   definition of 84–5
   in popularizations of quantum physics 81–2, 86–9, 97 (n10), 152–3, 164, 165
      anthropomorphic 17, 28, 83, 88–105
   in scientific discourse 83–6
mythic narrative 18, 107, 111–35, 164
   criticism of 116–19
   definition of 112–15

narrative 108–111, 139–40, 147, 165 *see also* mythic narrative
"New Age" movement
   and popular physics 31–5, 83, 94 (n7), 96, 100, 164
   definition of 31
   denouncement of 31, 33
   role in publishing "boom" 42
Newton, Sir Isaac 137
   popularizations of Newtonian theory 16, 21, 22
nuclear energy *see* atomic physics
nuclear physics *see* atomic physics
nuclear weapons *see* atomic weapons

Overbye, Dennis
   Lonely Hearts of the Cosmos 36
   recognition of narrative conventions in 120–21

Pagels, Heinz
   Cosmic Code, The 35, 95, 96
   denouncement of "New Age" movement in 33
Penrose, Roger 57, 125, 126
   The Emperor's New Mind 30, 94 (n7)
   readership of 12
   sales of 12
physics
   perceived pre-eminence of 6–7, 79
physics, quantum *see* quantum physics
physics popularizations 5–7, 10, 19–20
   of astronomy 21, 22, 27, 30
   of atomic physics 27–8
   of Big Bang theory 12, 30, 35–8, 107–31, 135
   criticisms of 32–3, 54–5, 57–8, 117–18
   definition of 11, 12
   and the "Hawking phenomenon" 132–34
   history of
      eighteenth and nineteenth centuries 20–23, 47
      twentieth century 23–39, 42, 47
      twenty-first century 46
   marketing of 38–9
   and "New Age" movement 31–5, 42, 83, 94 (n7), 96, 100, 164
   as mediators in "two cultures" debate 13, 16, 17, 18, 165
   and philosophy 25, 26, 28
   of quantum theory 27, 34, 81, 86–9
   of relativity theory 4, 24–5, 27–8, 31, 32, 34, 75, 91
   readership of 11–13, 20, 21, 23
   and the "Science Wars" 78–80
   spiritual beliefs in 22–3, 25, 31, 37–8, 152–3
   see also "New Age" movement
   stereotypes of scientists in 18, 137–61, 165

textual strategies of 3, 4, 17, 18, 21, 26–7, 28, 38, 81–2, 99, 100 (n14), 139–40, 141–3
  see also characterization, metaphor, narrative
  see also chaos theory; complexity theory; cosmology; quantum physics; science popularizations
Polkinghorne, John 37
  and Heisenberg's Uncertainty Principle 93
  popularizations by 37
  reaction to *A Brief History of Time* 49
  rejection of anthropocentrism in quantum theory 94
popular science
  definition of 7–10
popular science books
  accompanying lectures 21, 22
  accompanying television series 21, 30–31, 35
  role in information exchange 3–4
  see also science popularizations
public science
  definition of 8
public understanding of science (PUS)
  definition of 8–9

"quantum consciousness" 93–5, 98–102 see also metaphor
quantum physics 17
  Copenhagen interpretation of 87, 93
  and "New Age" movement 31–5, 42, 83, 94 (n7), 96, 100, 164
  popularizations of 4, 12, 24–5, 27–39, 75, 81, 91, 99
    impact on literature of 58
    metaphor in 17, 28, 81–82, 83, 86–105, 152, 164, 165
    see also physics popularizations
quantum mechanics 27, 34
relativity theory 19, 24, 118
Theory of Everything, the (TOE) 36, 37, 38, 129–30, 131, 133, 135

readership of science popularizations 11–13, 20, 21, 22, 23
  strategies used to target 13
relativist theories
  relativist-constructivist 67, 69, 70, 71, 72, 74 (n3), 75, 77
  of scientific knowledge 65, 66, 67–8, 163
  definition of 65
relativity 119, 125, 130
  General Theory of Relativity 24, 118
  popularizations of 4, 24–5, 27–8, 31, 32, 34, 75, 91
  impact on literature of 58
  Special Theory of Relativity 19
Russell, Bertrand 25
  ABC of Atoms 27, 90
  ABC of Relativity 24

Sagan, Carl 42
  Cosmos 21, 35
    analysis of rhetoric in 116–17
    use of mythic narrative in 121
  influence of 52
  popularization used in literary work 2
  Stephen Jay Gould's opinion of 35
science popularizations 5, 19–20 see also physics popularizations
  bestselling 7, 12, 17, 24, 25–6, 32, 35, 37, 38, 43–6, 47–8, 95, 107
  definition of 9–10, 11
  different models of 9–11
  history of 19, 20–39, 42, 46, 47
  literary criticism of 4–5, 14–16, 61
    theoretical approaches to 69–70, 73–5, 76–7, 164
  literary critics' reaction to 51–8, 95–6, 104–106, 108–09, 116–18, 164
  and literature 1–2, 3–5, 58
  as mediators in "two cultures" debate 13, 16, 18, 52, 53, 54, 57–8, 104–105, 163, 165
  post-war slump in publishing of 28–30
  as producers of knowledge 9, 10
    and use as information sources 82–3

publishing "boom" 1, 5, 16–17, 19, 31, 41–2, 107
   and the "Science Wars" 78
   and the "two cultures" debate 51–8
   characteristics of 42–47
   in response to Einstein's theories 24–8
   reception by literary community 42, 54, 55–6, 57–8
   role of *A Brief History of Time* in 47–51
   readership 11–13, 20, 21, 22, 23
   strategies used to target 13
   relationship with professional science writing 9–11
   textual strategies in 4, 17, 18, 21, 26–7, 28, 81–2, 141–2 *see also* characterization, metaphor, narrative
science studies 66
   reactions against 70–71, 72–4, 76, 78, 79, 81–2 *see also* "Science Wars"
   theoretical perspectives in 67–8, 69
   see also sociology of scientific knowledge (SSK)
"Science Wars" 1, 3, 17, 56, 107, 140, 156, 157, 161, 163–4 *passim*
   effect on popularizations 78–9
   and physics popularizations 78–80, 105
   and the publishing "boom" 78
   rhetorical tactics used in 73–7
scientific knowledge, theories of
   constructivist 10, 14, 61, 65, 66–7, 68, 69, 70, 71, 72, 73, 115, 163
   approach to language 83–4
   attacks on 71–3, 75–6, 107, 115–16, 119, 135, 156 *see also* "Science Wars"
   definition of 65, 67, 68
   relativist 65, 66, 67–8, 163
   relativist-constructivist 67, 69, 70, 71, 72, 74 (n3), 75, 77
scientific writing, language conventions in 69–70, 79, 83, 107, 138
scientists, image of 48, 54, 112, 113, 131–4, 137, 138, 139, 150, 151–5, 156, 159

female 157–61
   see also characterization; stereotypes of scientists
Smoot, George
   Wrinkles in Time 36
   publisher's advance for 50
Snow, C. P. 7
   "two cultures" debate 1, 3, 7, 17, 51, 54, 56, 163
   inadequacy of 2, 58, 61–4
social study of science (SSS) 66
sociology of scientific knowledge (SSK) 65–6, 67, 68, 78, 156–7
Sokal, Alan
   attack on science studies 72–4, 76, 79, 81–2
   reactions to attack 74, 164
Special Theory of Relativity *see* Einstein, Albert
Stannard, Russell
   Uncle Albert and the Quantum Quest 91
Stenger, Victor
   rejection of anthropocentrism in quantum theory 94
   Unconscious Quantum, The 94
   denouncement of "New Age" movement in 33
stereotypes of scientists 1, 18, 137–8, 140, 145, 149–50, 157, 160–61, 165
   gender 153 (n12), 157–60
   see also characterization
Stewart, Balfour *see Unseen Universe, The*
Stoppard, Tom 57
   popular science in work of 2–3, 58
   Arcadia 54
   Hapgood 86
Sullivan, J. W. N. 58
   Aspects of Science 25
   Atoms and Electrons 27
   observations on science popularizations 23, 31
   quantum theory 87
   Limitations of Science, The 25
   Three Men Discuss Relativity 24
superstring theory 124–5

Tait, Peter see *Unseen Universe, The*
textual strategies
　　use by science popularizers 3, 4, 17, 18, 21, 26–7, 28, 38, 81–2, 99, 100 (n14), 139–40, 141–3
　　see also characterization, metaphor, narrative
Theory of Everything, the (TOE) 36, 37, 38, 129–30, 131, 133, 135
theory of relativity *see* relativity
Thomson, William (Lord Kelvin) 21, 22
Tipler, Frank 23, 37
　　criticism of 117—18
　　publisher's advance for 50
"Two cultures" debate 1, 2, 3–5, 7, 17, 41, 42, 51, 54, 55, 56, 75, 107, 117, 118, 157, 163 *passim*
　　impact of science studies on 70–71
　　inadequacy of 2, 58, 61–4
　　and "New Age" movement 31
　　and publishing "boom" 51–8
　　role of popularizations as mediators in 13, 16, 17, 18, 52, 53, 54, 57–8, 104–105, 163, 165
　　see also "Science Wars"
Tyndall, John 16, 21
　　philosophy of 22, 23

Unseen Universe, The 22–3, 31
　　use of metaphor in 86

Waldrop, M. Mitchell
　　Complexity 17, 39, 108, 140–41(n7)
　　　　characterization of scientists in 18, 140–50, 160–61, 164–5
　　　　female characters in 157, 158, 159
　　　　narrative style 147–8
　　　　textual strategies in 39, 139–40, 141–3

Weinberg, Steven, 73
　　attacks on constructivist theory by 71, 74, 75, 78, 79
　　view of physicists' role in the "Science Wars" 79
　　Dreams of a Final Theory 36, 124, 129, 134, 137
　　　　emphasis on aesthetics in 119
　　First Three Minutes, The 17, 35, 107, 118 (n6)
　　　　female characters in 158
　　　　readership of 11–12
　　　　use of metaphor in 27
　　　　use of mythic narrative in 18, 107, 111, 112, 120, 121–3, 126–7, 128–9, 130, 131–2, 134, 164
　　　　public image of 131–2, 137, 153–4
Weldon, Fay
　　reaction to science popularizations 54, 55, 72
Whitehead, A. N.
　　Science and the Modern World 21, 25
Wolpert, Lewis 43
　　attacks on constructivist theory by 71–2, 74, 115, 156, 157
　　on the humanity of scientists 155–6, 165
　　and "two cultures" debate 54, 57

Zohar, Danah 158
　　popularizations and "New Age" movement 35, 94 (n7)
　　use of metaphor by 91–2
Zukav, Gary 32, 96
　　Dancing Wu Li Masters, The 17, 32–3, 42, 81, 95, 96
　　　　female characters in 158
　　　　use of metaphor in 83, 86, 95–105, 152–3, 164